Poultry: Anatomy, Breeding and Genetics

Poultry: Anatomy, Breeding and Genetics

Edited by:
Nada Ben Abdallah

Delve
PUBLISHING

www.delvepublishing.com

Poultry: Anatomy, Breeding and Genetics

Nada Ben Abdallah

Delve Publishing
2010 Winston Park Drive,
2nd Floor
Oakville, ON L6H 5R7
Canada
www.delvepublishing.com
Tel: 001-289-291-7705
001-905-616-2116
Fax: 001-289-291-7601
Email: orders@arclereducation.com

ISBN: 978-1-77361-248-5 (Hardcover)

ABOUT THE EDITOR

Nada graduated from the University of Zurich in 2010 and holds a PhD in neuroscience. Her research focuses on elucidating the processes of learning and memory as well as the affective processes and their underlying neural mechanisms. Her work so far involved normal and diseased animal models with a main interest in neurodegenerative diseases. She is currently working as a freelance researcher and science editor.

TABLE OF CONTENTS

SECTION 3: POULTRY BREEDING

LIST OF FIGURES

LIST OF TABLES

CHAPTER 6

CHAPTER 7

CHAPTER 8

CHAPTER 9

PREFACE

Poultry is one of the fastest growing and developing subsectors in agricultural industry. There has been unprecedented growth in production of poultry meat and egg attributed by proper genetic selection, breeding, housing, and feeding programs in this subsector.

Poultry rearing for egg and meat production has been the primary business for many people in different places around the world. Poultry has also been and still is a key animal product in most diets consumed worldwide. With various advancements achieved in preservation of techniques for poultry products and processed products, preference of consumers for poultry and poultry products are higher than ever. Basic information on the technology and science used in production and processing this crucial food commodity is significant to the work of academia, governments, and industry.

Several good and professionally prepared reference books are available, but the preference for any specific book greatly depends on the needs for the reader. Most of the poultry reference books are single-volume books, and some books only cover general topics, whereas others cover specific poultry topics.

This book has been prepared to provide essential information about poultry anatomy, breeding and genetics to all those student taking poultry management courses, veterinary professionals, teachers, poultry management staffs, and even various governmental staffs involved in agricultural sectors. The general science and various production practices presently in use in poultry industry have been characterised and well described so that the reader can acquire insights into the industry.

The book has been divided into four sections. Section one deal with general information on poultry rearing and it consist of only one chapter. Section two deals with background information on poultry anatomy and physiology and consists of two chapters. Section three deals with poultry breeding and consists of four chapters. Lastly, section four dealing with poultry genetics consists of three chapters.

In order to perform post-mortem on poultry and be able to evaluate various internal and external diseases of poultry that influence viability of a breeding program, students should understand the key anatomic components of poultry. This book is essential because it has relevant information on various anatomic systems of a bird-both internal and external (second section). It has also elaborated in details about various genetic principles and how they influence poultry breeding and production practices.

More emphases have been put in breeding and production of meat type chickens and egg producing poultry birds because these are the two key areas where most breeders and producer focus on. Specific areas tailored towards minimizing production costs while maximizing overall production have also received much of the attention, specifically on the third and fourth section of the book (poultry breeding and genetics).

Editor

SECTION-I
INTRODUCTION TO POULTRY

CHAPTER 1

INTRODUCTION TO POULTRY

The word poultry is used to refer to domesticated birds species mainly reared by human being to satisfy certain needs, mostly food. The most common types belong to the orders anseriformes (waterfowls such as geese and ducks), and galliformes (such as turkeys and chickens). Generally, the term poultry is used to refer to birds' meat. There are different types of poultry, but the following are the major and accepted species: turkeys, chickens, ducks, guinea fowl, goose, pigeon, ostrich, and pheasant. There are different uses for poultry in the modern days. One of the main uses of poultry is production of food. Most domesticated birds are reared for their meat and eggs. Most poultry species are consumed in the whole world and are not discriminated by religion. There are certain breeds reared for amusement such as ornament, pets, sports or games.

Some species like the ducks produce soft feather mainly used for stuffing winter jackets, blankets, and pillows. During earlier eras, the feathers of quail were used as writing tools. An ostrich skin on the other hand is commonly used in manufacturing leather items. Some cultures use the bones and meats of certain chickens' breeds (such as black-skinned Chinese silkies) for medicinal purposes. Other uses of poultry include instant source of cash income and for organic fertilizers.

DOMESTICATION AND POULTRY REARING

The idea of taming certain species of birds and rearing them for various purposes started early during human civilization era. Studies shows that human beings have being rearing tamed chickens as early as 6,000 B.C. Other species such as turkeys, geese and ducks were domesticated later. As human civilization progressed, most people adopted better ways of domesticating and rearing the birds. It was discovered that each poultry species have different characteristics, and these characteristics if managed in a better way could be passed on to the offspring. Most people preferred the best breeds that produced more eggs and meat, while others preferred and emphasized on their aggressiveness or physical appearance.

As poultry farmers became enlightened and gratified about the potential and benefits of each species, they came up with different breeds and varieties based on unique variants in their flocks. For instance, as the chickens were being moved from southern Asia mainly through human migration, conquest, and trade to other geographical areas, most people became more interested in rearing and propagation of the birds especially those that had distinct characteristics. Hence, after several centuries, different breeds and varieties of turkeys, chickens, ducks, and geese have been recognized and developed by farmers in different parts of the world. Fascinatingly, up to now, most of the domesticated chickens reared in southern Asia are similar to the jungle fowls that existed thousands of years ago.

Economic significance of poultry has really grown. This has been highly attributed by advancement and industrialization of different societies. Farmers are continuing enlightening themselves on how to improve their poultry production through adopting poultry species with superior traits. Poultry breeds with poor meat and eggs production are been scraped out of the rearing systems. Today more farmers are able to obtain more high quality breeds through hybridization.

POULTRY SPECIES

Chicken

This is one of the most and widely reared poultry specie. It was first domesticated about 7,400 years ago. Chickens belong to the *galliformes order*. It is believed that this poultry specie originated from grey jungle fowl (*Gallus sonneratii*) which was hybridised with the red jungle fowl (*Gallus gallus*). Jungle fowl primarily dwell in highly dense vegetation is social groups of about four females, one male, and a few chicks. Chickens spend most of the day time on the ground foraging for food and roost at night. About 75% of their day time is used in foraging for the food. Apart from foraging, chickens also use a considerable period of time dust-bathing and preening. Chickens have different breeding seasons with summer and spring being the best seasons.

Due to technological advancement and awareness, different techniques have been adopted to help in improving the quality and breeding traits of modern chickens. Some of the physical changes that have resulted from domestication and hybridisation of chicken include enhanced growth rate, increased body weight for the adult birds, earlier sexual maturity, simplified plumage, larger and more frequent eggs production.

Figure 1.1: Modern chicken breed.

Source: *http://www.poultryhub.org/species/commercial-poultry/chicken/*

Chickens can be classified based on the type of breeds available. A breed can simply be defined as a group of birds that have the ability to reproduce offspring that look alike. In other words this means that the offspring share certain traits with their parents. However, there might be some differences in one or two traits, for instance the colour of the feather and shape of the combs. Worldwide, there are mainly two distinct types of chickens breeds kept for commercial purposes, these include meat type and the eggs type with several breeds in each type. The meat type chickens (commonly referred to as broilers) and egg types (also known as layers) both are kept for meat and eggs production, and their differences are on how efficient and effective they produce these commercial commodities.

Most pure breeds are likely to disappear due to over hybridisation of the few available poultry breeds by the commercial poultry industry. There are different types of poultry exhibitions, clubs and shows formed to protect and maintain the standards of poultry.

There are more meat type breeds than the egg type breeds. Most meat type breeds have the capacity to produce more than one kilogram of live-weight from less than two kilograms of poultry feeds given to them. On the other hand, some of the best egg type chicken breeds can produce more than 350 eggs annually.

Duck

Domestication of duck was first done about 4,000 years ago. Most ducks are kept for eggs, meat and feather production. Most ducks products are still popular especially in Asian regions. There are different ducks breeds reared worldwide.

The first breed type is the Pekin duck (*Anas domesticus*). Some of the physical characteristics of this breed include broad body, orange beak and feet, and whitish-creamy feathers. It was first domesticated and developed in china during Mongols period. Because of its rapid growth rate, Pekin duck is mostly used for meat production. Studies show that this breed is able to attain a body weight of about 3.5 kg within a period of 6 weeks. Apart from meat production, Pekin duck is also used for egg production in some places. In other words, it used as a multi-purpose bird in poultry farming. It is also important to note that Pekin duck usually exhibits little broody behaviours, a trait that is apparent in other types of duck breeds.

Figure 1.2: First ducks to be domesticated 4,000 years ago in China.

Source: *http://www.poultryhub.org/species/commercial-poultry/duck/*

Muscovy duck (*Cairina moschata*) is another breed of duck that is believed to have originated from South America. Genetically, this breed is more like the geese than the duck breeds. Some of the distinctive characteristics of this duck breed include bare face, red caruncles around the beak and eyes, they also prefer foraging for grass in the field, and egg incubation period for Muscovy duck is usually 35 days. The growth rate of Muscovy duck is believed to be lower than other duck breeds, where it takes about 10–12 weeks to achieve its market live weight. Market weight of the males is usually 45% more heavily than the females' market weight. In other words, males can attain about 5 kilograms at slaughter while the female can reach up to 3 kilograms at slaughter. Because of this difference in growth rate of males and females, most poultry farmers have adopted a single sex rearing strategy. Other traits of Muscovy duck include high broodiness and low egg production rate. Despite these negative traits of the duck, Muscovy duck is still one of the best breed commonly used worldwide because of its lean meat.

Indian runner belonging to the *Anas platyrhynchos* order is another duck breed. This breed is known for its prolific egg producing ability. Indian runners are capable of producing about 365 eggs in 1 year. It is believed to have originated from south Asia. It has a tall and upright stance.

Khaki Campbell belonging to the *Anas platyrhynchos* order is used as a dual purpose breed because of its ability to produce eggs and good meat. One of the main and distinct traits is its khaki colouration. It was first developed in England in 19th century through crossing breed between the Indian runner and Rouen duck. This breed has the ability to produce more than 340 eggs in 1 year.

Mule also known as Mulard duck was developed through crossing breeding *Anas platyrhynchos* and *Cairina moschata*. This type of breed is mainly sterile hybrid, reared predominantly for fat liver and meat production. This duck has the best lean meat that is desired by most health sensible people.

In 2013, the total duck meat production was about 4.37 tonnes, with Asian region accounting for more than 80% of the production. Actually, Asia is considered as the best and leading duck producer in the whole world. China being the first country to domesticate duck about 4,000 years ago is one of the key producers of duck meat. Statistically, China produces about 3 million tonnes of duck meat every year, which represent about 85% of meat production in the whole Asian region and about 60% of duck meat production globally. Although there are different types of ducks used for meat production, there are three main breeds commonly used in commercial meat production; these include Mule duck, Muscovy duck, and Pekin duck.

Although most non-Asian citizens do not consume duck eggs as compared to eggs of other poultry breeds such as chicken, some countries like Indonesia and Vietnam keep ducks primarily for table eggs production. Local duck breeds in these two countries are usually reared in flocks comprising of about 100 birds, which are free ranged on rice fields to forage for their feeds and follow rice harvest in order to collect insects, fallen grains, fish, frogs, and other aquatic animals.

Figure 1.3: Young duckling

Source: *http://www.poultryhub.org/species/commercial-poultry/duck/*

Young ducklings are usually incubated and hatched within a period of 4 weeks. Unlike chickens that take about 3 weeks, most ducklings are usually brooded for about 2 weeks. Ducks are known for their first growth rate and both sexes are usually grown together. Like chickens, duck are also reared in open sided vermin-proofed poultry houses with deep litter cover and generously spaced ground floor. Due to their tendency of drinking plentiful amount of water, they usually excrete watery droppings and spill drinking water if it is offered in troughs. Because of this, the litter in their house usually get soaked quickly, thus requiring frequent changing.

Most ducks are usually grown to about 7 weeks of age and males gain about 100 grams per days with about 2:1 feed conversion ratio. Duck breast meat is the most crucial part of duck production, which accounts for about 25% of the carcass weight. Another important part of the duck meat is the thigh meat which accounts for 30% of the carcass weight.

Most duck species are known for their resistant to various poultry diseases; however, they are known to be the best carrier of different infectious poultry diseases. For this reason therefore, they are usually reared away from other poultry in commercial and farmyards. Most diseases that affect ducks are usually non-infectious and result from various breeding programs aimed at enhancing production. Legs problems are one of the predominant issues in ducks production. Anatomically, ducks are designed for propulsion through the water than bearing their weight on dry land. They are also likely to develop legs problems if they are kept in a cemented house or in wire mesh. Some of the common legs problems include broken and fractured bones and foot sores. Other reported diseases that affect ducks include *Rimerella anatipestifer, Salmonella,* and *E. coli* outbreaks.

Geese

There are more than sixty different types of geesc breeds in the world and most of them are commonly found in the eastern parts of Europe. The first domestication took place about 6,000 years ago in China, whereas in Egypt took place about 3,000 years ago. Geese belong to the *Anser genus*, and are believed to be progenies of the wild swan geese *Anser cygnoides* and *greylag goose* (*Anser anser*) commonly found in most parts of Asian region. Some of the best geese reared for meat production include the brown and small white Chinese geese, African geese, grey Toulouse geese, and the small Embden geese.

Figure 1.4: The brown Chinese goose domesticated about 6,000 years ago

Source: *http://www.poultryhub.org/species/commercial-poultry/goose/*

The brown and small white Chinese geese are the most common in the southeast parts of Asia and they usually weigh about 4.5–5.5 kilograms. The female geese can lay up to a hundred eggs in one season. Unlike other common European geese, most Chinese geese usually have a larger knob which is located at the base of its bill.

Most geese in different areas of the world are usually reared for meat production, egg production, fatty liver, and feather and down production. Studies have shown that geese meat is one of the best and contain essential fats used for different medicinal purposes. Depending on the type of breed, incubation period for hatching geese eggs range from thirty to thirty five days. About ninety five percent of all geese meat consumed worldwide (2.5 million tonnes) come from China. Other high geese meat producers include Egypt, Ukraine, and Hungary. About nine hundred and fifty tonnes of the meat is produced in United States of America.

In the year 2010, geese meat production was estimated to around 2.5 million tonnes globally, and this accounted for about 3% of all poultry meat production in that year. China has always been the best and leading country in geese meat production accounting about 2.4 million tonnes in 2010.

Figure 1.5: Young geese chicks.

Source: *http://www.poultryhub.org/species/commercial-poultry/goose/*

Geese production is usually different from that of chickens. Most geese usually prefer mating in water where the female should be served with a male goose about 1 month before the onset of a breeding season. Generally, female geese usually start to lay eggs when they attain an age of about 40 to 50 weeks. They lay about 30 to 45 eggs weighing about 120 to 170 grams per egg. Fertility and hatchability of geese is usually lower than other poultry species (usually 60 goslings in every 100 eggs set). For this reason therefore, geese farmers are highly advised to adopt a moderate poultry feed restriction for their geese 2 months before breeding season in order to increase their fertility and hatchability. Just like the ducks, young goslings usually have a high growth rate. With proper management they can attain a body weight of about 2 kilograms within a short period of time (4 to 5 weeks of age) and full maturity is attained once they are about 2 years old.

Apart from high growth rate, most geese breeds are highly resistant to different poultry infections, have long life spans, they are hardy, and can survive for a long period of time foraging without other supplementary poultry feeds. This is because of their ability to extract proteins from their foraged herbage and their capacity to digest fibre obtained from their diets. Geese are usually reared in small groups with one male for every four female geese. They are mainly free ranged but sometimes are reared in pens with straw beddings or slatted floors.

Most geese breeds are versatile in a way. Before, they were used for controlling weeds in various crops in United States. In some places they are also used as homestead guard dogs. Geese are also reared for their livers that are mainly used for foie gras production.

Turkeys

This poultry specie is believed to have originated from North America. Just like the chickens, turkeys belong to the order *Galliformes*. There are usually two distinctive kinds of turkeys; the Ocellated Central America turkey (*M. ocellata*) and the wild North America turkey (*M. gallopavo*). Annually, turkeys' meat production is estimated to about 6 million tonnes, which 60% coming from North America which is the native home for most of the modern turkeys. Actually, there are still some untamed turkeys (*Melagris ocellata*) still found in North America. Although turkeys were domesticated about 500 years ago, their meat was exploited in the 20th century. Before, turkeys were reared for various poultry exhibition shows and plumage.

Most female turkeys (hens) are usually smaller and generally less colourful than the male turkeys (stag/tom). The wingspan of most adult turkeys is about 1.5–1.8 meters and the males have fleshy appendage hanging from their beaks. It takes about 4–5 months for a pure turkey breed to grow to maturity. Their lifespan is estimated to about 10 years. In terms of sex differences, female turkeys usually take about 16 weeks to full maturity and attain an average weight of 3.5–7 kilograms. The male turkeys on the other hand take about 19 weeks to fully maturity attaining about 7–11 kilograms. With time a male turkey (tom) can actually grow up to 18 kilograms if it is well managed. Based on a female's 6-month egg laying cycle, studies have shown that a breeder tom turkey has the ability to produce up to 2,000 poults. Commercially reared turkey hens can produce about 50 live chicks (poults) annually. Incubation period for turkeys' eggs takes about 4 weeks to hatch. High mortality is usually recorded on early stages of turkeys' development. One of the major contributing factors to high mortality is poor hatchery hygiene that results in navel and yolk sac infections.

Figure 1.6: Black and White turkey breeds

Source: *http://www.poultryhub.org/species/commercial-poultry/turkey/*

Most commercially reared turkeys for meat production are usually hybridised, usually between more than one breed with different characteristics. Hybrid turkeys have a fast growth rate and able to convert feed to bodyweight more efficiently than the pure turkey breeds. Under perfect rearing conditions with the right poultry feeds ration, the current white hybrid turkey has the ability to attain about 6 kilograms in body weight within a period of 10 weeks of age, with feed conversion ratio (the amount of poultry feeds consumed to body weight gained by the bird) of about 2:1. Most commercially reared turkeys are hybridised to specifically increase the amount of meat on their thighs and breasts. White turkeys are usually the most preferred by many farmers because they do not have nasty stain spots on their skins when plucked.

Most adult male turkeys (toms) usually have a carbuncle (a skin hanging from the beak) that changes its colour during courtship. Turkey meat is one of the best and commonly used in the US especially during various cerebrations such as Christmas and thanksgiving. Breast meat accounts for about 75% of the entire carcass value while thighs meat account for 10%. The fat content on breast meat is usually about 1.5%. Turkey rearing especially during the early stages is usually difficult as compared to the chickens. The slaughter weight of most turkeys will depend on the end use of the bird (table meat, restaurant meat, or for further processing), the price of turkeys meat is very competitive in places like America.

Due to hybridisation, most modern male turkeys (toms) have gained more breast meat, thus making mating difficult. Because of this reason, most turkey hens are inseminated artificially. Because of fast growth rate and the focus on the breast meat, most turkeys are faced with different leg problems resulting from their excess body weight placing more pressure on their legs. Tibial dyschondroplasia is one of the commonly reported skeletal disorders are a result of fast growth rate.

Table 1.1: Illustrating other different poultry breeds that have being domesticated by human beings

Common names	Scientific names of the birds	Place of origin and the period of initial domestication	Primary uses of the birds
Quails	Also referred to as *Coturnix japonica*	Originated from japan and first domestication was done around 11th century	Mainly reared for meat and egg production
Pigeon	Scientifically referred to as *Columba livia*	Not clear	Mainly used as pets and messengers. In some place used for meat (especially from young squab)
Guinea fowl	Also referred to as *Numida meleagris*	Originated from Africa. First domestication was done about 1,500 B.C.	Mainly reared for eggs and meat production
Pheasant	Also called *Phaseanus colchicus*	Originated from Africa. First domestication was done during early roman civilization.	Mainly kept for meat and eggs production.
Ostrich	Also referred to as *Struthio camelus*	Originated from Africa. First domestication took place in 19th century.	Mainly reared for skin and meat.

Source: https://www.thoughtco.com/the-domestication-history-of-chickens-170653

PROBLEMS AND CONSTRAINTS FACED IN POULTRY FARMING

Just like any other type of farming, poultry farming is faced with several problems and constraints that undesirably affect its comparative effectiveness and competitiveness in the international market. Some of these problems include climate problems, economic and trade policies, infrastructure facilities, and poultry infections among other challenges.

Climate

Most poultry species reared in farms are usually warm-blooded (homeotherms) animals. They can maintain their body temperature. However, this is only achievable if the temperatures are within certain limits. If these limits are exceeded, their productivity is usually reduced drastically and sometimes may result to death. For instance, the ideal temperature for a laying chicken usually ranges from 13 to 24 °C. In the intensive commercially reared poultry ambient temperature in poultry house is usually increased by the heat generated by the birds. This condition usually results to heat stress on the birds and high mortality rates may also be recorded. To reduce these problems, most poultry farmers, especially in commercial industries usually install cooling systems, which also add to the already high poultry production costs.

High temperature and adequate humidity in poultry house also form favourable environment for the growth of various poultry infectious organisms such as bacteria, viruses, parasites, and worms. This condition makes it difficult to actually control and prevent various infections that affect poultry in the farms. Eggs and meat production usually deteriorate at alarming rates during humid and warm conditions. This results to huge losses if amicable strategies are not adopted to curb the problem.

Economic and Trade Policies

With intensity and commercialization of poultry farming, farmers are more sensitive to various economic and trade policies of a country. Since 2000, increased interest rates, devolution of currencies, and liberalization of trade among others have negatively affected poultry industry in one way or other. Because most international poultry farmers depend on imported production inputs such as poultry feedstuffs, breeding stocks, veterinary products, and other poultry equipment, the local cost for poultry farming is usually higher. Additionally, due to low tariffs rates and the removal of various quantitative restrictions under liberalization trade policies, poultry meat and egg importation has dramatically increased in various countries, thus leading to huge loses. This is because most local poultry producers are not able to get maximum profits from their locally produced products due to increase dependence of imported poultry products.

Inadequate Poultry Infrastructure Facilities

Just like any other enterprise, poultry farming highly depends on public network transportation, communication and other infrastructures. Poultry farmers also need to be insured and secured from various lawless elements. Additionally, there is the need to provide adequate accessibility to various extension services and other credit facilities. Due to lack of these services, most poultry farmers are reluctant in undertaking various poultry ventures, and those that engage into poultry farming usually end up encountering huge losses, high production costs, and poor productivity.

POULTRY PRODUCTION SYSTEMS

Poultry can be reared using different production and management systems. There are several factors that determine the type of poultry production system to use in a farm. These include the type and breed of the poultry, the level of both inputs and outputs, mortality rate of the poultry, producer type, production purposes, broodiness length, growth rate of the birds, and the number of poultry birds been reared. There are three key poultry production systems commonly used in modern farming. These include intensive poultry production system, semi-intensive poultry production system, and free-range poultry production system.

Intensive Poultry Production System

This is one of the most effective systems used in production of poultry commonly used in various farms today. It is also one of the most expensive systems because it requires more inputs in terms of poultry feeds, health, breeds, housing among other inputs. It is mainly used in chicken rearing and the number of the birds involved are usually high (not less than 200 birds). It is used for the layers and broiler productions. There are usually categorized into two main types; that are deep litter poultry production system and cage poultry production system.

For the deep litter poultry production system, the birds are reared on the floor, usually littered with about 5–10 cm thickness litter. Some of the material used for litter include dry hay, coffee pulp, teff straw, and sow dust. To prevent occurrence of various diseases in the house, the litters should be kept clean and dry. All drinkers and feeders should be raised off the littered floor to help in preventing dampness in the house. Deep litter poultry system can be used for both broiler and layers production.

Figure 1.7: Layer chickens been raised in a deep-litter house system. Source: http://northernhomestead.com/the-deep-chicken-litter-method/

Some of the advantages of using deep litter poultry rearing system include proper poultry accommodation, prompt culling of the unproductive flocks, predators and disease control, high egg production, and good record keeping. This system also uses small piece of land. It is also one of the best and economical poultry rearing system. Effectiveness on poultry feeding is enhanced, high level of poultry production supervision, requires less labour, the value of manure is also increased, and it is also easy to automate various production processes.

Deep litter offer the best isolation against heat loss during cold seasons. Studies have shown that egg productions in layers are usually higher if they are reared in a deep litter system than other poultry keeping systems. Some of the disadvantages of using deep litter system include high capital investment, high susceptibility in case of a disease outbreak, and high rate of cannibalism. If this system is poorly managed, accumulation and liberation of ammonia, problems of dirty eggs, and infections may result. Tick and worm infestation cases may be high in a deep litter system than other poultry production systems. This system is also not the best for areas with high humidity due to high rate of disease occurrence and spread.

For the cage poultry production system, birds are reared in nests or cages and it is one of the best systems for the layers. The eggs laid by the birds are purposely used for food because there are usually no cocks to fertilize the eggs. Most chickens reared through this system usually get their nutrients through supplementation because they do not have access to free ranging. Despite high initial investment capital incurred in this system, most farmers enjoy good profits.

Figure 1.8: Cage poultry production system

Source: *http://www.esdaw.eu/battery-cage.html*

Using cage system, more birds can be reared as compared to other poultry production systems. The system also requires less labour to maintain various activities of rearing and managing the birds. Monitoring poultry production is also easier using this system. This means a farmer can identify poor breeds and cull them appropriately, thus saving production cost. Problems of infections and parasites can be reduced using this system. Some of the disadvantages of this poultry rearing system include high installation cost for the cages, breeding difficulties unless insemination is done artificially, paralysis or cage layer fatigue issues especially if it is not attended earlier, high initial cost for investing the system, there are more broken eggs in this system compared to other system such as deep litter system, and the system involves high labour intensive.

Semi-Intensive Poultry Production System

This system is better than the free range poultry production system. This is because it uses various poultry inputs such as vaccines and poultry feeds supplements. The birds are usually accommodated in a small house. This house is used as resting place for the birds especially during the night. The house has a door on one side to allow free movement of the birds in and out of the house during the day. There is also a fence sounding the house to prevent the birds. Since the birds are not able to forage for feeds, supplementation with proteins and other energy feeds is recommended.

Some of the advantages of suing this system are that it facilitates easier control of production operations, good record keeping, economical use of the available land, and the birds are well protected against predators. On the other hand, the system requires high initial cost especially in fencing the land. Most birds are also exposed to different poultry infections especially if the compound is not clean and dry.

Free-Range Poultry Production System

This system of poultry production is mainly practised in the rural areas. The main objective for this system is to rear poultry mainly for household consumption and providing extra household income. The free-range system accounts for about 95% poultry production especially in various developing countries. It is actually not very profitable because the birds reared are usually not market oriented. Most indigenous poultry birds are raised using this system and depend mainly on the local available feed materials. Health service and other management practises are usually minimal for this type of poultry production system. Young chicks depend on the mother/hen for care and protection, and because of this, young birds are usually predisposed to high rates of mortality, exposure to poultry disease, and predators. Some of the main sources of feeds include the worms obtained from scavenging, cereals, legumes, and sometimes are supplemented with some feeds especially where there is shortage of feeds. Production is usually less 100 eggs per bird annually.

Figure 1.9: Showing free-ranged chickens.

Source: *http://www.farmerstrend.co.ke/majority-free-range-kienyeji-poultry-farmers-wrong/*

Some of the advantages of using this system in poultry production include minimal poultry infection with parasites especially if there is enough space available for the birds, labour input is usually minimal, chickens are usually healthy because of exercising in the open air, and management of the poultry farm is minimal. On the other hand, most birds are usually vulnerable to predators and unfavourable weather conditions. Most eggs are usually lost because the birds are not accustomed to laying the eggs in nests. There is also high poultry disease transmission thus causing high mortalities. It is also important to note that most birds reared using these systems are usually less productive as compared to the other poultry production systems.

POULTRY HOUSING

Housing is crucial for poultry in order protect the birds against predators, provide shelter, prevent theft, and protect against harsh weather (sun, rain, dropping night temperature, and extreme cold winds). A comfortable or suitable poultry house is significant for enhancing production and convenience of the poultry farmer. There are certain preconditions that should be looked in to when constructing a poultry house. First, the site of the house should be dry and secure. This means that the soil should be able to absorb ground water freely. It is also important to construct poultry house in a place where there is easier access to water, transportation, electricity and other necessary infrastructures.

There are different types of poultry house system that can be construction. The kind of poultry house to adopt is determined by availability of constructing materials, the type of poultry been reared, and the weather condition of the place among other things. It is also important to consider the cost of building the poultry house. This is because some poultry housing systems are more expensive than others. Durability and immediate gain of the house to the poultry is also important considerations to always put in place. There are five main type of housing system used in poultry farming. These include a poultry house with all side open; a poultry house with only one side open; a poultry house with a guard; poultry house with all side closed, and finally a poultry house with deep pit.

For a poultry business to be successful, there are certain standard of poultry production equipment that must be met. The type of equipment used inside poultry house are determined by the age of the birds, productivity status of the birds, and the type of poultry breed been reared. Some of the main equipment used in poultry farming includes the following.

Feeders-whether been used in the intensive, semi intensive, or free range poultry production system, feeder are usually the same. It is important to provide enough feeders in the house so that all the birds can feed at the same time. These feeders should also be kept clean at all times in order to prevent poultry disease occurrence and spread. Poultry feeders can either be made at home or can be commercially obtained from the market. A good feeder should avoid feed wastage and contamination of poultry feeds. It should also be easier to clean, strong, durable and easier to fill.

Drinker-just like the feeder; same drinkers can be used either in intensive, semi intensive, or free range poultry production system. They should also be kept clean at all times to prevent contamination and disease outbreaks in the poultry house. There are different types of drinkers available in the market and the kind of drinker to use should be determined by the age of the birds, size of flock, and the productivity of the birds been reared. Some of the characteristics of a good poultry drinker are that it should give fresh, clean, and enough water to the poultry at all times. A good drinker must be stable, durable and strong. It is also important to select a drinker that is easier to clean and fill.

Laying nest-this are very important equipment or structures in the house of poultry. They provide suitable place for the egg laying birds. Laying nest should be provided in the house prior to the breeding season so that the birds can have enough time for adoption before they start laying the eggs. Laying nests are important because they help to identify unproductive birds in the farm. They also prevent cracked and dirty eggs, especially if the eggs are collected regularly. Laying nest should not be built outside poultry house in order to prevent thieves and predators. When building laying nests, it is highly advisable to consider placing them above the ground. They should be enough for all the birds and big enough for the birds to always feel comfortable when laying the eggs.

Table 1.2: Illustrating feeding space, water space, housing space, and laying nest requirements for chickens at different ages

Feeding and watering space requirement in cm/bird					
Feeder/ Drinker type	Chicks 0-8 weeks	Layer growers 8-18 weeks	Broilers	Layers	Breeders
Long feeder	7	9	5	12	18
Round feeder	3	4	2	5	7
Long drinker	1.5	2	1	2.5	3.5
Round drinker	1.5	2	1	2.5	3.5
Housing space requirement (number of birds in 1m^2)					
Housing type					
Full slats			15	7	-
2/3 slats, 1/3 litter			-	6	4
1/2 slats, ½ litter			-	5	-
Full litter			10	4	3
Laying nests requirement					
Types of nests					
Individual nest (hens/nest)				7	5
Communal nests (hens/m^2)				50-75	30-50

Source: *http://www.hyline.com/aspx/redbook/redbook.aspx?s=2&p=46*

Litter- there are several advantages of providing litter in poultry house. These include providing comfort and warm to the birds especially during cold seasons, house drying through absorbing waste material from poultry excretes, and control breeding and transmission of disease causing organisms. There are different materials used for poultry house litters. These include sow dust, dried hay or grass, by-product of coffee plant, and cereals straw (such as wheat straw, teff straw, barley straw, etc.).

Good quality littering material should be able to absorb moisture in the house well, it should not cause harm or injuries to the birds, should have a moisture level of less than 15%, should easily decompose to form manure, should be evenly spread in the house of poultry, should be toxic-free, and should not cause pollution of the dust.

Proper management on poultry house litter is always necessary to prevent contamination. The litter should be place about three days prior to introduction of the birds in the house. It is also advisable to spray the litter on a weekly basis with appropriate disinfectants in order to prevent and control pests from breeding in the house. Regular removal and replacement of the litter is necessary where it get contaminated, dirty, or soak.

POULTRY HEALTH

There are different ways of maintaining good health for the poultry in the farm. First, it is advisable to always provide well-balanced poultry feeds rations to the birds at all times. There are different benefits achieved if the birds are provided with well-balanced ration, these include maintaining good health, promoting steady poultry growth rate, prevention and controlling susceptibility to various poultry diseases, and increasing diseases resistance to poultry diseases. A good balanced poultry ration should contain all essential nutrients such as carbohydrates, proteins, vitamins, and minerals.

Apart from providing well-balanced rations to the poultry, there is the need to control and prevent occurrence of various diseases in poultry farm. This is important in order to maintain good health of the poultry. There are various general methods of preventing and controlling occurrence of disease in poultry. These include vaccination against various viral diseases, keeping poultry house clean at all times to control disease causing pest and parasites, disinfections of the house, and using various prophylaxis and medicines for disease preventive measures in poultry.

SECTION-II
POULTRY ANATOMY

CHAPTER 2

ANATOMY AND PHYSIOLOGY OF POULTRY-I

The difference between poultry (birds) and other mammals based on the structural features is that birds have feathers, they do not have teeth, they lay eggs, they are able to float and fly, and they use only one orifice for waste excretion from their bodies. The word anatomy is derived from a Greek word "to cut up." Poultry anatomy refers to the scientific structure of body and organs of a bird. Physiology on the other hand is the science dealing with the function of various living organisms and their organs.

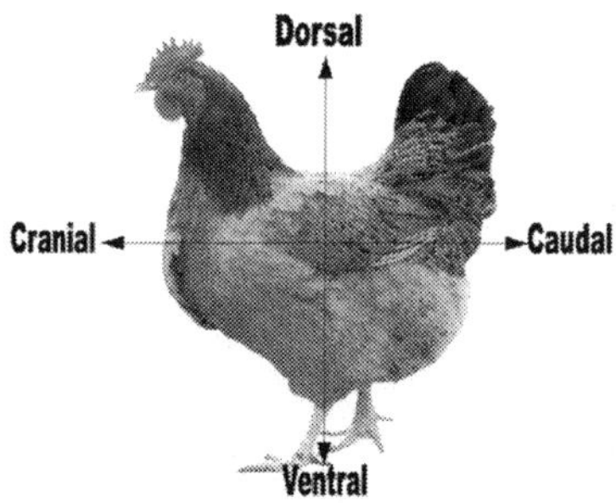

Figure 2.1: Anatomical directional terms on a bird's body.

Source: *https://www.slideshare.net/Ibnalbeetar/poultry-anatomyandphysiologypres*

There are several terms that are commonly used in describing the locations on the body of birds. This include dorsal (which refers to the upper part of the birds body), ventral (referring to the abdominal and lower parts of a bird's body), cranial or anterior (refers to the head or the front part of a bird's body), and caudal or posterior (refer to the rear or tail of a bird).

Just like any other animal, the body of a bird is sub divided into several parts. These include integumentary system, thermoregulation and respiratory system, skeletal system, digestive system, circulatory system, urinary system, reproductive system, endocrine system, excretory system, immune system, metabolic system, important sensory organs and nervous system, and muscular system.

INTEGUMENTARY SYSTEM

This is the out surface of birds' body. It comprises the feathers, skin of the birds, and the appendages (beak and claws). The skin mainly covers the body of the birds and has specious glands on the outer canal of the ear and preen gland on the tail base. The preen gland is used by the birds to secrete crucial oil for conditioning the feathers. The integumentary system is a crucial part in the body of poultry. It protects the birds from various dangerous conditions. Some of the main functions of integumentary system include offering protection on internal body organs and system of the birds from various external environmental conditions. In other words, this system provides protection against physical injuries and infectious microorganisms. It is also an important body system that helps in regulating and insulating the birds' body from variable environmental temperatures. Most nervous endings are found on integumentary body system to help the bird be aware of potentially risky situations. Integumentary body system helps in pigmentation of the birds. Another very important function of this system is to help in conversion of vitamin D especially when the birds are exposed to the sunlight.

The Skin

The skin of a bird has three major cells layers or tissues. These include the epidermis layer (which forms the outermost part or tissues of the birds skin), the corium or dermis (comprising of the middle tissues of the skins), and the hypodermis (which is the innermost tissues or layer of the skin). The skin of poultry is made up of several areas. These areas are modified in order to help the bird execute different functions. These areas include the feathered skin, the lower feet and legs covered with scaly skin, the horny and hard areas of toenails and beak, plantar (foot pad) and the wattle and comb skin.

The feathered areas of the birds skin usually comprises of special tracts. One of these areas is called pterylae-which is the areas that promote growth of the feather. The other area is called apteria- which is the areas that do not support growth of the feathers.

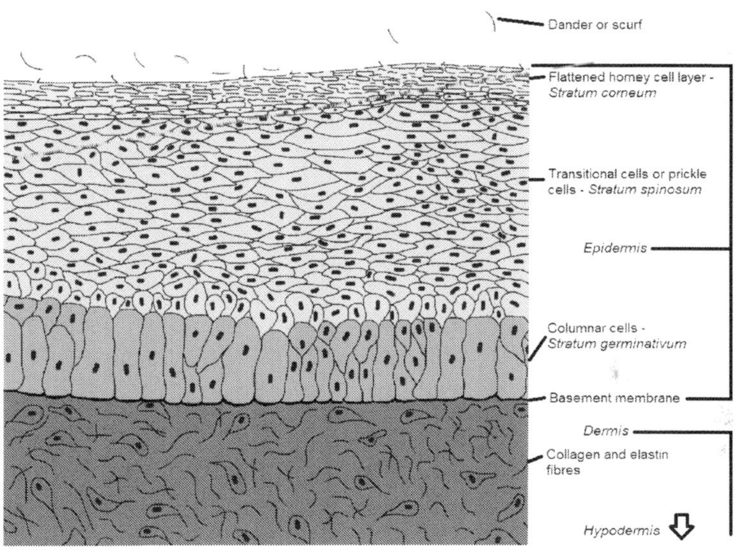

Figure 2.2: The structural layout of poultry skin.

Source: *http://www.poultryhub.org/physiology/body-systems/integumentary-surface-of-the-bird/*

Poultry skin structure-the skin of the birds is made up of two distinctive tissues. These include the epidermis cellular stratified epithelium, which comprises of the horny and hard outer skin layer and forms the feathers. The other one is the dermis connective tissues and hypodermis with widely spaced cells. Different types of poultry skins have different amounts of special compound such as keratin, elastin, and collagen.

The epidermis tissue of poultry skin has three distinctive layers. These include the horny and flattened cell layer that form the outermost layer of the birds skin; the transitional skin layer that joins the inner layer to the outer layer of the skin; the columnar cell skin layer that forms the innermost skin layer of the bird. This layer also forms the germinative layer that facilitates formation of various epidermis cells.

The dermis-The dermis of birds' skin is usually thin with microscopic and uniform structures. The key component that form this part of a birds skin is the collagen, but there are small amounts of elastin. Hypodermis has fewer cells which are loosely arranged than in the dermis. Most of the birds' fats are stored in this part in special adipose cells (fat tissues). Hypodermis also has air spaces that are usually connected to various respiratory system air sacs. This helps in enhancing the ability of poultry to fly.

Most cells on the skin surface are continuously replaced through mitosis processes in the epidermal germinative layer. This means that as the birds outer skin wear off or is lost, it is usually replaced by the underneath cell layer. The tissue lost on the outer surface of the birds skin usually form a very crucial part of dander or scurf, which act as diet to various parasites such as mites and lice living on the surface of birds' skin. The key functions of epidermis involves keratinisation processes which include-manufacturing or synthesis of keratin (which offer the skin its resistance and strength to normal wear and tears) and dead cells (which become the horny and hard cells layer on the skin).

The skin also has several nerve endings that are located mainly on the outer layer of the skin dermis. These nerve endings provide the birds with protection from various harmful conditions through making the birds aware of their presence. The dermis layer also has connections of numerous blood vessels that help in transportation of essential nutrients, hormones, respiratory gases, and other essential elements passed through diffusion into the epidermis inner cells.

Skin of feathered areas: Most parts of the birds are usually covered with feathers. These feathers usually offer normal protection to these areas. The skin on the feathered areas is usually thinner than other uncovered areas. The skin on the thighs and wings of the birds is more closely joined to various underlying skin tissues than over the other parts of the body. The epidermis in these areas is usually twelve cells thick and the horny outer layer being five cells think. The transitional layer on the other hand is about two cells think and four to six cells think on the germinative layer.

Skin of the feet and legs: The lower parts of birds legs is usually covered by a scaly skin which is attached to the epidermis. Some poultry breeds have their legs covered by feathers that grow on the scale margins. The scales on the legs of poultry are believed to be associated to reptilian ancestors of most birds. They are also rich in keratin.

Skin of planter or the foot pad: The skin underneath the feet and toes of poultry birds is usually specialized in order to help in withstanding the density of the birds' weight when standing and various abrasions caused when the birds' feet come into contact with injurious surfaces as it move about. Although the skin on these parts is thick and hard, it is usually flexible.

Skins of the beak: The beaks of most poultry birds are made of two key parts- these include the internal bone skeleton and a highly keratinised skin covering the bony skeleton. The beaks epidermis usually develop a thick horny skin layer known as "rhampotheca," which is made of flattened sheets of cells that are firmly attached to each other. The beaks of hatching chicks are usually made up of toughly hardened keratin in order to help the chicks break out of the shell.

Skin of the combs, wattles, and claws: Most poultry birds' claws and toes are highly keratinized. The root of claws epidermis helps in formation of flattened keratinised cells columns which are firmly stuck together. They form the accessory sexual epidermal appendages on the neck and head of the birds. They result from secretion of certain sex organ hormones of both male and female birds when they are functional. The skin of combs and wattles is usually brightly red-coloured due to dense supply of blood vessels. During hot weather, most birds usually deep these special organs in cool water in order to cool them.

Preen Gland

This is a special pea-sized structure which is usually developed from the skin epidermis of the birds. It is usually located at the tails base on the last vertebra of the birds. It usually releases fatty secretions via a duct opening on a small pimple or papilla on skin surface. These secretions are made from degenerated cells and are holocrine. They contain both saponifiable and non-saponifiable lipids, lecithin, and other cell degradation products. These secretions are similar to secretions of sebaceous glands of mammals and it is believed to have similar functions. These glands secretions are collected by birds' beaks especially when preening their feathers, and are used for conditioning the feathers. For the aquatic birds, these gland secretions provide water proofing on the feathers of the birds.

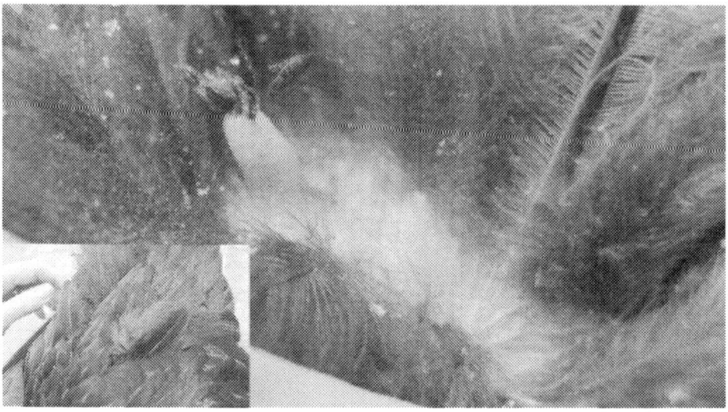

Figure 2.3: Showing the preen gland at the base of the chicken's tail.

Source: *http://www.the-chicken-chick.com/2013/03/chicken-feather-loss-cannibalism-causes.html*

Feathers

Most birds' feathers are usually confined to specific tracts of the birds known as pterylae. The areas without feathers are referred to as apteria. Most feather follicles are usually attached to smooth muscles. During cold weather, the feathers are erected in order to increase the volume of air trapped. This helps in providing more insulation to the birds. This process is also used by the birds during hot weather in order to release hot air in between the feathers. There are five major types of feathers in most poultry fowls. These include the contour feathers, the bristle feathers, the plumules, filoplumes, and the down (chick).

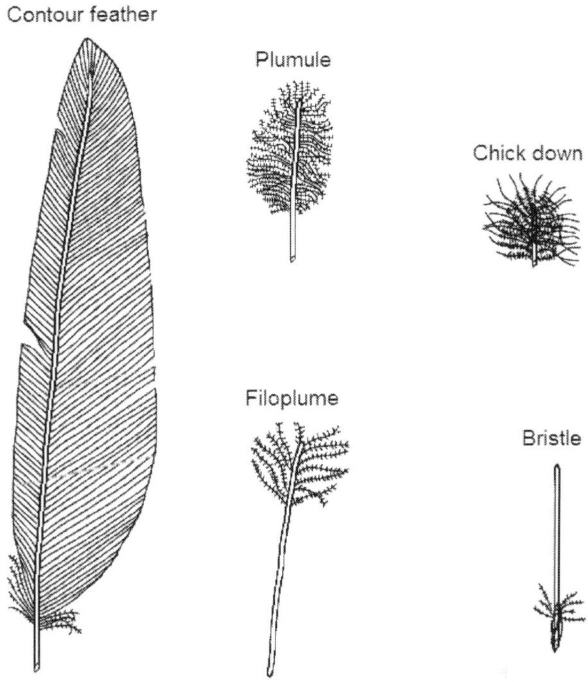

Figure 2.4: Five major types of feathers in poultry.

Source: *http://www.poultryhub.org/physiology/body-systems/integumentary-surface-of-the-bird/*

Contour feathers: They are made of the rachis or the shaft; the vexillum, which slope diagonally towards the tips; the quill or calamus, which are usually transparent, rounded, hollow, and have several conical scales in their inner parts; the inferior umbilicus, the hole located on the shaft at the junction of vanes and shaft; the hyporachis or aftershaft, which is minute feather-like tuft; and the rachis, which is an elastic and four-sided shaft that runs the vane length.

Most contour feathers are usually different in shape and size. The main feathers of the wings (known as remiges) have broader posterior parts of the vane than the anterior parts (posterior part is the rear or back while anterior parts are the front parts). Contour feathers are made of the primary and secondary flight feather, and are usually separated by axial feathers which appear smaller in size. Most primary flight father are found mainly on the outer or trailing edge of the wings while the secondary flight feathers are much closer to the body. Each wing of an adult poultry bird has about ten primary flight fathers. The large feathers located on the tail of the birds are known as reticles. There are also other special contour feathers called coverts used to cover rectrices and remiges bases (e.g., tail coverts and wing coverts).

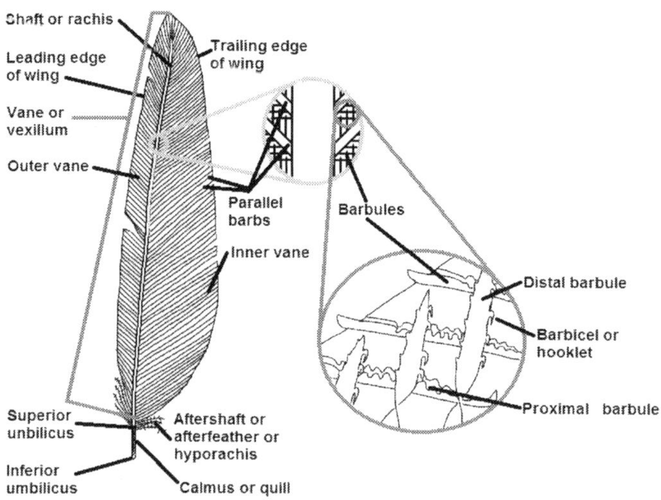

Figure 2.5: Illustrating different sections of contour feather.

Source: *http://www.poultryhub.org/physiology/body-systems/integumentary-surface-of-the-bird/*

Plumules: Most plumules are mainly located underneath the contour feathers. They form a soft and downy undercoat. Plumules have shorter shaft with free, radiating barbules and bards. Most of them do not have interlocking barbicels or hooklets. These feathers play very crucial role of insulating the birds during cold weather by trapping air and holding it still.

Filoplumes: These are small and hair-like feathers with barbules and barbs confined to the apex. Filoplumes are usually troublesome small feathers during live poultry processing into poultry meat due to their difficulties in removing them.

Chick down: These feathers are commonly seen on newly hatched chicks. They are usually fluffy and closely look like plumules with numerous radiating barbs. After the first moulting of a birds, most juvenile feather usually emerge from the feather follicle in order to replace the fluff. The juvenile feathers on the outermost parts look like contour feathers of an adult bird. With time, the juvenile feathers are replaced by final plumage. There are several moulting sessions before formation of adult contour feathers. The number of moulting is usually specific for each type of feather follicle, but this may be predetermined by pterylae.

Bristles: Most bristle feathers are located around the eyes and mouth areas of the poultry birds. They are primarily believed to help in birds' sensory abilities and protecting various sensitive parts of the birds. Bristles can be compared to the eyelashes of the mammals.

The Feather Follicle

Biologically, feathers are derivatives of keratinised epidermis which are linked to epidermal scales. Embryonic follicles develop as dome-shaped epidermis projections over raised pimple or papilla of the dermis. These dome-shaped projections later sink into the skin by downward growth rims. The rim base commonly referred to as collar continue showing some rapid cell divisions during follicle formation. Most of the cells that help in formation of the feather through differential growth are usually arranged into special structural patterns of rachis and calamus, barbs, barbicels, and barbules.

During feather development and formation stages, the dermis is usually densely vascularised (dense blood vessels to supply essential nutrients necessary for the formation of feathers). After keratinisation of feathers cells, the dermal follicles dries up forming the pulp commonly found inside the calamus. The final stage of feather formation involves splitting of feather sheath, resulting to formation of new feather unfurls. Development of new feather follicle usually stops after hatching period is complete.

Feathers Colouration

The colour of the feather (brown, black, and reddish brown) is caused by the granular pigment (melanin) which is deposited in the feather cells that have been keratinised. Melanin is made from various non-essential amino acid (such as tyrosine) secreted by special pigmentation cells known as melanocytes. During embryonic period, the pigment cells enter the dermis and congregate underneath the follicles of the feathers. After keratinisation, melanin granules usually move into the cells of epidermis. Melanin colouration may be reddish or blackish (phaemelanin).

Apart from melanin, carotenoids are also used in colouration of birds' skins and feathers. Carotenoids are the orange and yellow lipid pigmentation elements usually found in free fats of various poultry dietary ingredients. Carotenoids are responsible for the yellow pigments on the legs, beak, feathers, and skins of poultry. The yellow colour of the egg yolk is also formed by these carotenoids. Some of the sources of carotenoids include lucerne leaves, carrots, and the maize.

The Physiology of Moulting

All poultry feather follicles usually undergo a certain recurring changes of growth (anagen) and resting period (telogen). Most growth stages of feather follicle occur before old plumage moulting. After growth of the feather follicle is complete, feathers are usually attached to the resting follicle by their calamus or quills.

Figure 2.6: Moulting physiology in chicken.

Source: *https://www.pinterest.com/pin/278941770647141502/*

After the telogen period, the germinal epidermal cells usually enter an anagen stage. In this stage, quills are developed from old feathers and are pushed out of the follicle canal, but still remain attached to the new feathers horny sheath. After sometimes these sheath splits to expose newly developed feathers. This cause the old feathers to fall away inevitably (mounting). Feather moulting is a dependent process that occurs after the growth of new feathers on the skin of the birds. Most poultry birds can initiate anagen phase through plucking of feathers on their body.

Feather growth is determined by various sex-linked genes. Based on the kinds of sex-linked genes, a bird can be described as late or early feathering. Moulting process of various poultry fowls can be induced by secretion of certain hormones in the body of the birds. One of the major poultry hormones that play an important role in moulting process is called thyroxin, which is secreted in the thyroid gland. This hormone helps in stimulating growth, differentiating, and feathers patterning through increasing metabolic activities of epidermal follicle cells. Although chicken down is not predetermined by secretion of this hormone, development and growth of feathers at juvenile and adult stages depends on availability of this hormone. The growth of wings feathers is not influenced by thyroxine. Feathers will stop growing except those on the wings if thyroid glands are removed from the birds' body. Female poultry birds are more sensitive to thyroxine than the male poultry.

High level of female sex hormone (oestrogen) may also retard formation of feather when the feather follicles are in the telogen stage. Both oestrogen and testosterone play a significant role in development and formation feathers. These hormones are actually responsible for the differences that occur between females and males. For instance, most male birds will develop female-like feathers if secretion of androgens is limited.

Light pattern also play a crucial part in development and formation of feather follicle of birds. Studies have shown that light have significant influence on the pituitary gland, which helps in production of thyroxine by the thyroid gland and gonadotropins by testes and ovaries. Long day length will stimulate the production of thyroxine and gonadotropins while short day length will cause a decrease in the production of these hormones.

SKELETAL SYSTEM

Skeletal system forms the framework that offer protection and support to various organs, tissues, and other body systems of poultry fowls. The bones of birds are usually homologous to those of other animals. They have evolved in order to enhance birds' ability to fly. Other modifications include fusion of various vertebral sections to enhance rigidity necessary for flight; have pneumatic bones where respiratory system air sacs are connected to long bones cavity; have sternum that offers a large surface are necessary for flight; most birds have a long neck which act as shock absorber for various brain tissues especially when landing.

Vertebrae

The vertebral column of poultry skeletal system has thirty nine different bones categorized in to five main groups. These include the coccygeal vertebrae, cervical vertebrae, thoracic vertebrae, lumbar vertebrae, and sacral vertebrae. There is a specific vertebral formula commonly used in describing the vertebral column of the birds. The formula is as follows: C13, T7, L+S14 & C5 = 39. This formula illustrates the total number of vertebrae in each column section.

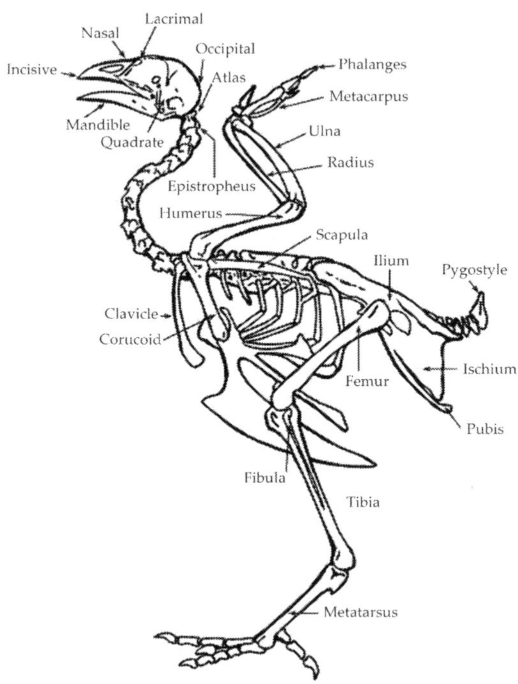

Figure 2.7: The vertebral structure of poultry skeletal system.

Source: *http://gallusdissection.blogspot.co.ke/p/skeleton-and-bone.html*

Atlas: The atlas is located underneath the skull of the birds. Compared to other vertebral column, atlas has different structure. It is much smaller than other vertebral column and has a ring-like shape with deep articulation cavity. They also have another specialized column called axis. The axis of the birds' skeleton is short and passes via the ring-like atlas. This vertebra helps the birds to turn their head freely.

Lumbar and thoracic vertebrae: Thoracic column comprises of seven vertebrae. Most ribs of the poultry are found on this vertebrae column. Thoracic vertebrae also offer processes for muscles attachments. The second to fifth thoracic vertebrae are usually fused together to offer the necessary structural strength required by most poultry birds for flight. The coccygeal, thoracic, sacral and lumbar vertebrae are usually considered as one group because they are closely fused together. The last vertebrae of coccygeal are also fused together, thus forming pygostyle that offer strong foundation for most tail feathers.

Ribs: An adult poultry fowl has seven pairs of ribs projecting from the thoracic vertebrae. All ribs pairs except the first and second do not reach the sternum (the main attachment point). The third and sixth pairs divided into two main segments-which are vertebral segment and sternal segment. All the ribs apart from the first rib pair and the seventh rib pair have uncinate (meaning bent or hooked) processes projecting backward over the outer surface of next ribs. All the ribs are connected to each other by ligaments, and this provides more strength to the thoracic cavity.

Sternum: This is also called the breastbone. It has the most complex shape of all the other vertebrae. It is usually termed as quadrilateral, which comprises of curved plate projecting from the middle of caudal and cranial board. Caudal medial projection also referred to as metasternum forms the longest projection made of tall plate-like sternal or ridge crest. This plate runs from the front part of the ventral vertebrae surface. Sternal crest also referred to as keel bone offers suitable attachment for various flight muscles such as the subcontractors and pectoralis muscles.

Skull: The skull of most poultry birds is usually divided into two main region; which are the conical facial region and rounded cranium region. The difference between these two regions is brought about by the two large orbits into which the eyes of the bird fit. These orbits are separated by the septum which is made of ethmoid bones and sphenoid bones. Based on external cranial appearance, the skull appears to have a large brain capacity. Other distinctive features commonly found in the skull include the eardrum, the jugular veins, carotid arteries, and the spinal cord. The location of the eardrum on the skull is represented by the tympanic cavities that are located at the back. The openings located at the cranium base illustrate the direct connection between the spinal cord and brain. Jugular veins and the carotid arteries are presented by the lateral openings of the occipital bone. Beneath the skull is the hyoid bone that forms the tongues framework. This bone has three key parts; which are entoglossal section, the basi-hyal section, and the third section. Both the entoglossal section and basi-hyal section are joined together by a flexible joint.

Wings or forelimbs: The skeletal structure of the wings or forelimbs resembles that of the theropod dinosaurs. They are essentially modified to perform special tasks such as feeding and flight. The wing skeleton is divided into two main sections. These include the shoulder girdle comprising the scapula, clavicle, and the coracoid. The other section is referred to as the wing which comprises of the humerus (upper arm), radius and ulna (forearm), and the carpus, metacarpus, and the digits (the hand).

Legs: The skeleton of the legs can be subdivided into two main sections; hipbones which comprises of ileum, ischium, and the pin bones or pubis. The other section is the leg which comprises of the thighbone or the femur, the tibia, and the foot (tarsus, metatarsus, and toes).

Poultry Bone Formation

The skeleton structure of most animals is essential for various reasons. This includes providing protection and support to the tissue and other body systems. Bones are made of living tissues and are usually affected by various natural stresses. Chemically, most poultry bones are made of essential minerals such as calcium and phosphorous. Most metabolic processes and activities of the bones are usually continuous and always changing. Most of the new bone tissues are deposited on the bones by small cells known as osteoblasts, while the existing bones are resorb by the large cells known as the osteoclasts. There are other special cells referred to as osteocytes which help in maintaining the structure of the newly formed bones.

Bone formation is influenced by the amount of vitamin D consumed by the birds and the amount of certain hormones secreted in the body of a bird. Such hormones include the parathyroid hormones, growth hormones, androgenic hormones, oestrogenic hormones, and the calcitonin. Deficiency of vitamin D in the diet of poultry always put most birds at risk of various bone related complications. This is because poultry and other animals require this essential mineral for the assimilation processes of calcium in the bone. If poultry lacks vitamin D in its diet, they may develop certain conditions like rickets and gout.

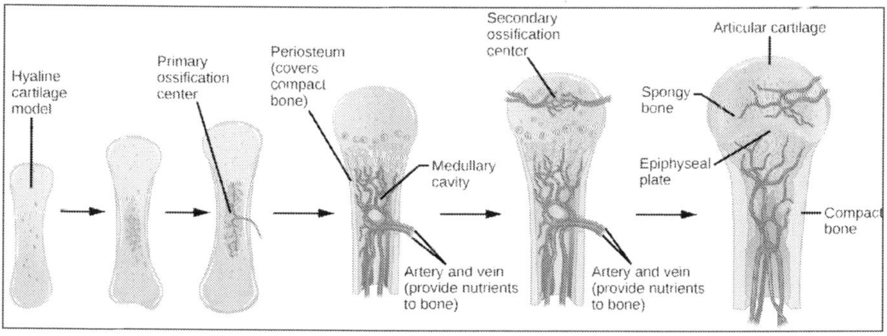

Figure 2.8: Bone formation process in poultry.

Source: *https://opentextbc.ca/biology/chapter/19-2-bone/*

Formation of the bones in poultry usually undergoes three main stages; which are membranous or prechondral stage, cartilaginous or chondral stage, and the ossification stage. Most poultry bone usually undergoes through the condral stage, but some like the skull bones usually omit this stage. The membranous stage of bone formation in poultry usually takes place in eggs during the embryonic development stage.

Secretion of chondroblast cells initiate formation of the cartilage. These cartilages are then hardened using the ossification processes when essential mineral such as calcium carbonates are assimilated in the bones. Through ossification processes, the long bones of the poultry are lengthened. These bones have a hollow cavity in the middle which is extensively filled with bone marrow and a dense network of air sacs.

DIGESTIVE SYSTEM

Compared to other animals such as cattle, goats, and sheep, poultry have the simplest digestive system. Because of this simplicity in the digestive structure of poultry, it is highly advisable to provide high quality and easy to digest poultry feeds. This is essentially significant if the birds are expected to have maximum performance in terms of eggs and meat production. This part is going to discuss various parts of poultry digestive structure and their functions. Digestive system is responsible for food ingestion, breakdown of this food into absorbable nutrients, and excretion of the waste out of the body. The digestive system of poultry is made up of several parts.

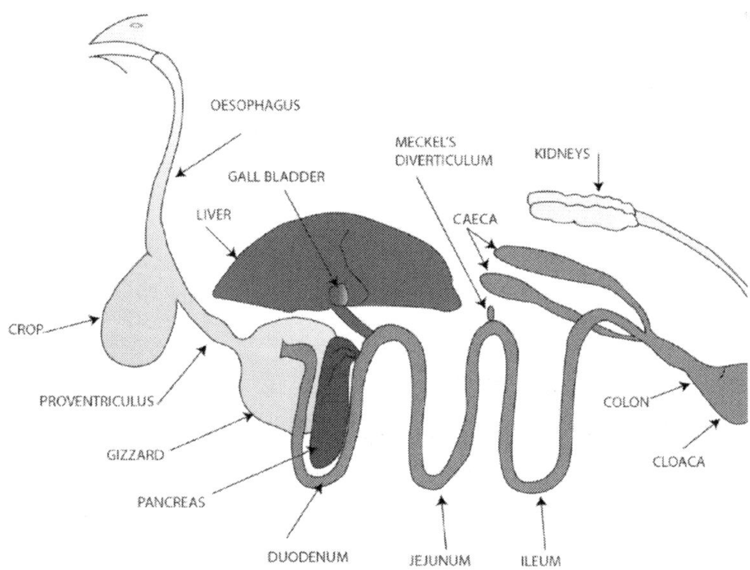

Figure 2.9: The digestive system of poultry.

Source: *http://www.poultryhub.org/physiology/body-systems/digestive-system/*

Alimentary Canal

This is the longest tube-like organ which starts in the mouth (beak) of the bird and end in the cloaca or vent in the abdominal region of the bird. This tube has several muscles running along the tubes to help in movement of ingested feeds through contraction. There are also some linings of mucous membranes along the tube. There are also several digestive glands located in different parts of alimentary canal. These glands are essential in secreting important digestive juices that helps in breakdown of the ingested foods. Digested food is also absorbed along the walls of the alimentary canal into the blood circulatory system and transported to various parts of poultry body for assimilation. The remaining waste is excreted from the bird's body through the vent or cloaca.

Mouth structure: Poultry do not have cheeks and lips; instead they have beak, a tongue, and taste buds. The mouth of poultry birds also has a hard palate which forms the mouth roof. This hard palate has a long and narrow slit that connect with the nasal cavity of the birds. This palate is also made of five key rows of conical papillae pointing backwardly in the mouth of the birds. There are also several salivary glands that help in initiating digestion of food in the mouth cavity through secretions.

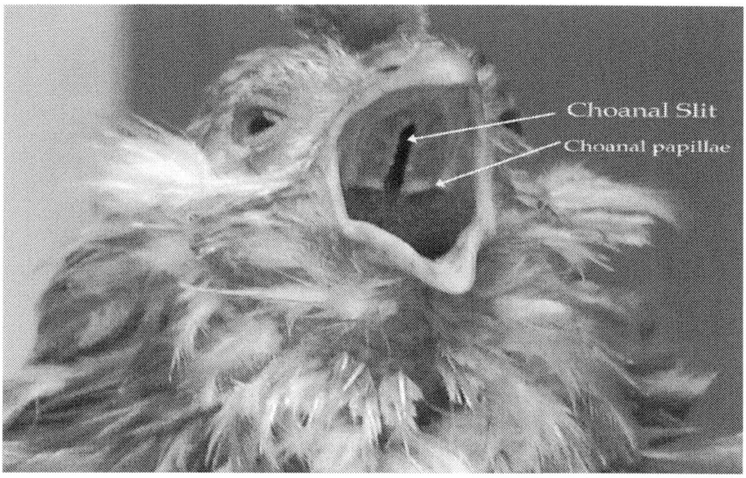

Figure 2.10: The interior of upper mandible on a healthy chicken.

Source: *https://www.backyardchickens.com/threads/mysterious-beak-growth.923843/*

Salivary glands: Salivary glands are located on the hard palate of the birds. There are different salivary glands in the digestive system of poultry. These include (1) the maxillary, which is located at the mouth roof, (2) the palatine salivary glands, which are found on nasal opening of the mouth roof, (3) the posterior sub-mandibular glands and cricoarytenoid glands, which are located near the glottis, (4) the lingual gland found in the birds' tongue, and (5) the apheno-pteryoid glands and another small salivary gland located in the mouth's angle.

Pharynx and tongue: Poultry has a continuous pharynx that starts at the mouth. Both the mouth and the pharynx of the birds together form a combined cavity commonly referred to as the oropharynx. The tongue of the birds on the other side is long and conform the beaks shape in which it functions. It has a horny and thick epithelium on the tip. On the posterior end of the tongue, there are transverse horny and simple row of papillae. The support to the tongue is provided by a special bone called the hyoid bone. Most of the mucous glands are commonly found on the tongue. Although it is not yet proven, many people believe that the tongue of birds has special taste buds on the surface. The tongue has two key functions in the digestive system of poultry; these are to receive food particles in the mouth and to push the ingested food in to the oesophagus.

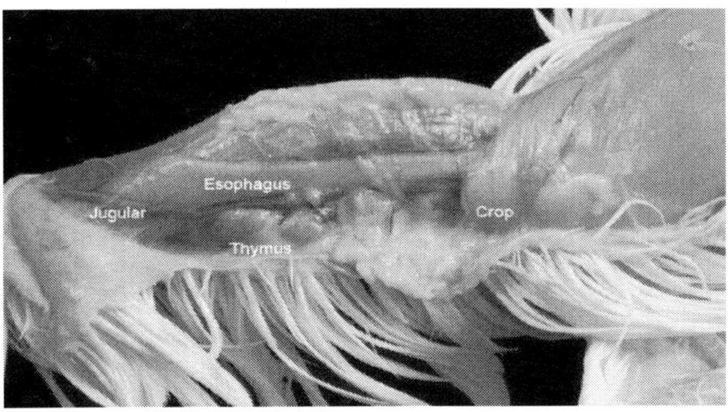

Figure 2.11: An overview of Oesophagus, crop and proventriculus in poultry.

Source: http://articles.extension.org/pages/65376/avian-digestive-system

Oesophagus, crop, and proventriculus: Oesophagus is a wide tube that has the ability to stretch to allow movement of the food down in the alimentary canal. This digestive section of a bird connects the crop to the mouth region of the bird. The crop on the other hand is a huge oesophagus dilation located next to the thoracic cavity. Its main function is to temporally hold the food for a short period of time before commencement of further digestion along alimentary canal. As the oesophagus enters the thoracic cavity, it forms another important organ called the proventriculus, which is sometimes referred to as glandular stomach. The structure of the crop is similar to that of the oesophagus except that most birds do not have digestive glands in this organ.

Proventriculus is relatively tubular and small organ along the alimentary canal of the birds. The wall of this organ is usually thick and have five key layers; the mucous membrane, thick layer of glandular tissues, areolar layer of tissues, the muscle layer and the outer serous membrane. This organ has several glands on the think walls that secrete different digestive enzymes and juices that help in breakdown of the food into absorbable nutrients. There are also several tubular glands located on the mucous membrane of the organ. These glands secrete hydrochloric acid which also helps in the digestion of the food.

Gizzard: This is a muscular stomach of the bird located just after the proventriculus, next to the liver and the lobes. It has a rounded and flattened shape similar to that of convex lens. This organ is covered by tendinous glistering layer of tissues which thicken at the centre and become thinner at the edges. The inner surface is made of horny tissues which appear creamy-coloured and raised in ridges. The outer layer on the other hand is made of tough and strong red muscles. These muscles are essential because they help in preventing damaging effects of various insoluble materials in the gizzard as the grinding of the food is taking place. Gizzards contain numerous hard objects such as grit and gravel that help in food disintegration, which is the main function of gizzards. Gizzards have special glands on the wall that secrete certain keratinised liquids which are usually passed to the horny linings surface of the gizzard to replace the worn out tissues.

Small intestine: It starts at the ending of the gizzard and ends where the colon and caeca meets. It is made up of three parts, which are the duodenum, the jejunum, and the ileum. Most of food digestion and absorption of nutrients usually take place in small intestine. Some of the essential functions of small intestine include secreting numerous enzymes that are used in the digestion processes of the food, it also provide the best medium for much of food digestions, and helps in the absorption processes of nutrients for assimilation in the body of the birds.

Duodenum: It forms an elongated loop which is about twenty centimetres long. This organ has lymphoid tissues that are usually located in the corium. This tissue is useful in the collection of lymph. It also has a bile duct that secretes some bile juice used in breakdown of the food. Another important organ located between the arms of the duodenum loop is the pancreas. This organ also secretes digestive juice useful in breakdown of the food. Both jejunum and ileum together form a long canal of about 120 cm long. The structural formation of these two sections of small intestine is usually similar to that of duodenum except that there are shorter villi on the walls, less lymphoid tissue, and suspended in the mesentery.

Large intestine: Both the small and large intestine have a similar structural formation except that large intestine is usually slightly smaller in size. It appears in a straight line running below the vertebrae and end at the vent or cloaca. Young birds have bursa of fabricius just before the cloaca and this section usually disappear when the birds reach the age of 1 year.

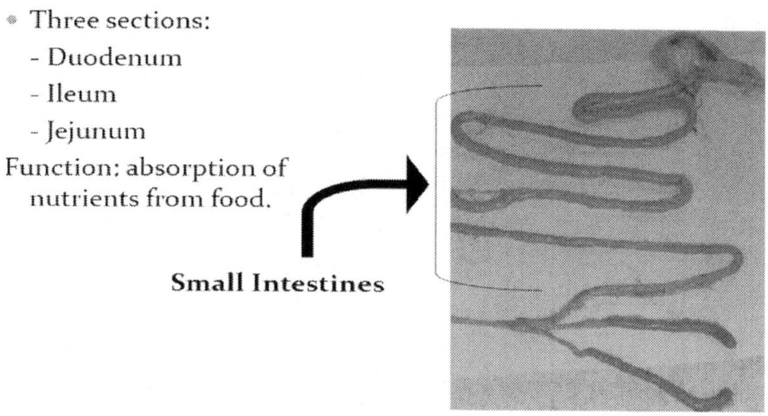

Figure 2.12: Structural representation of poultry small intestine.

Source: *http://www.dummies.com/home-garden/hobby-farming/raising-chickens/necropsying-a-chicken-the-internal-organs/*

Caeca: It forms the blind pouches in most adult poultry which is about 15–20 centimeters long. There are two caeca in a bird which are attached to the small intestine. There are three key sections of each caecum. These include the wide blind apex which has fairly think walls, the middle section with thin walls, and the narrow base which has thick walls. Structurally, caeca is made of circular muscle tissues, serous membranes, longitudinal muscles on the outer side, and mucous membrane of inner muscles.

Cloaca: It corresponds to the termination point of the large intestine. It forms a tubular cavity that opens on the outside of the birds' body. Structurally, cloaca resembles most parts of the intestine except most of its muscular mucosa part usually disappears as its open to the vent of the bird. Cloaca is subdivided into different chamber, and each chamber is separated from the other by a constriction. These chambers include the copradaeum which is an extension of the rectum and colon, the urodaeum which is the middle section where genital and urcter ducts open, the proctodaeum which form the part of intestine that open to the vent exterior. Young bird usually have a dorsal opening (bursa of fabricius) instead of proctodaeum.

Liver

This is one of the most important digestive organs that is found below and behind the heart and is closely related to the spleen and proventriculus of the birds. It is a bi-lobed chocolate coloured organ. In adult poultry, this organ weighs about fifty grams. The liver of birds is covered with a thin capsulated membrane.

The liver is connected to a special duct of the gall bladder. Liver has several functions in the body of a bird. These include formation of bile, metabolism of various nutrients, synthesis and destruction of various red blood cells, and production of fibrinogen and plasma proteins. Studies have also shown that liver is responsible for the storage various fats, glycogens, and some of the fat-soluble vitamins such as vitamin A. Liver also plays a very significant role in detoxification of various toxic substances ingested by most birds.

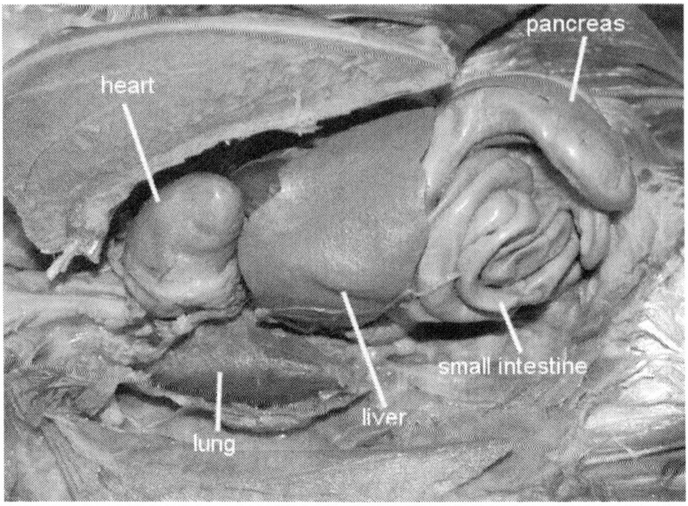

Figure 2.13: Illustrating major organs involved in digestive system of poultry.

Source: *http://people.eku.edu/ritchisong/birddigestion.html*

Pancreas

This organ is located between the two arms of the duodenum loop. It has three ducts connected to the duodenum distal end to facilitate movement of digestive secretions (pancreatic juices) into the alimentary canal. One of the main functions of the pancreas includes production of pancreatic juice which helps in digestion of ingested food. This organ is also responsible for the production of glucagon and insulin hormones which help in carbohydrate metabolism processes.

Poultry Digestion Information

The secretory and digestive processes in the body of poultry birds are influenced by the pattern of food intake and the movement of this food along the alimentary canal. Because of high and active metabolic rate processes in the digestive tract of the birds, it is necessary to continuously supply poultry with enough and well-balanced feeds. The eating behaviour of most birds widely differ from one fowl to the other, this means that some birds are fond of eating small amount of feeds over short interval while others consume large amount of feeds over a wider range of intervals.

Birds have a special organ called crop which hold undigested foods over a period of time and after sometimes it is then passed to other alimentary canal organs for farther digestion processes. Most birds consume feeds during the day light at an average interval of about fifteen minutes. Large feeds intake usually happen early in the morning and late in the evening. There are several factors that affect or influence food intake in poultry. These include the age and live weight of the birds, environmental conditions such as temperature and humidity, food energy content, the texture of the food ingested and the level of various nutrients in poultry feeds, the health status of the birds, egg production of the birds, and the quality and quantity of drinking water among other factors.

Digestion of Poultry Feeds

Food digestion in poultry is usually a long process that starts in the mouth and end in the intestines. Most poultry feeds are usually broken in to smaller digestive particles in the gizzard. There are different substances that help in breakdown of this food along the alimentary canal. These substances include enzymes, hydrochloric and gastric acid, and other juices secreted by various glands. Amylase enzyme secreted in the salivary gland in the mouth of the birds helps in breakdown of various complex carbohydrates. Most food breakdown takes place in the glandular stomach with the help of various enzymes and juices such as enzyme pepsin which helps in digestion of protein, gastric juice and pancreatic juices released in proventriculus and pancreas respectively. Apart from the pancreatic juice, the pancreas also produces insulin and sodium bicarbonate which help in maintaining the levels of sugar in the blood and the right pH level in the intestines respectively.

Most of the nutrients in the digested food are absorbed in the small intestines through special features called villi. Absorbed nutrients are then transported to various parts of the birds through blood streams for assimilation purposes. The remaining undigested food materials are then mixed with secreted urine and eliminated from the birds' body through the cloaca as faeces. On average, an adult egg laying bird produces about 100–150 grams of faeces. Water content in fresh faecal waste is usually 75%.

CHAPTER 3

ANATOMY AND PHYSIOLOGY OF POULTRY-II

CIRCULATORY SYSTEM

The circulatory system plays a very important role in the body of various poultry birds. Some of the essential functions include the transportation of absorbed food nutrients from the intestines to other organs of the body for assimilation purposes. It is also responsible for the removal of metabolic waste materials that accumulate in various body organs of the birds. Other materials transported by the circulatory system include medications, blood, antibodies that helps in fight various poultry infections, and worn-out cell residues.

Some of the key organs involved in poultry circulatory system include the heart, lymph vessels, blood vessels, bone marrow, and spleen. Vascular and blood system are the first system that develop in an embryo because of transportation of nutritional elements to various areas needed. There are different fluids that form part of circulatory system in the body of a bird. These fluids are categorised into two different types, namely the circulatory fluids (comprising blood and lymph fluids that move around the body) and the static body fluids (which are usually stationary in the body of the birds). Some of the static fluids include cell sap or the intracellular fluid, tissue fluid or intercellular/interstitial fluids, joints synovial fluids, vitreous and aqueous humor fluids, and finally perilymph and endolymph fluids of the ear (cochlea).

The heart acts as the main organ in the circulatory system for pumping the blood around the body of poultry. Some of the key vessels used in the transportation of the blood include the arteries, veins, capillaries, and lymph vessels. Once the arteries leave the heart, they are subdivided into smaller arteries necessary for supplying blood to various organs and cells around the body. Once these arteries reach various tissues, they are again subdivided into capillaries that have much smaller and thinner walls (usually one cell thick). This facilitates easier movement of nutrients from the capillaries in to various tissues of the body. As the capillaries leave various body tissues of the birds, they are re-joined to form other important vessels called veins which again re-join to form pulmonary veins or vena cava which move the blood back to the heart.

There is another special system called lymphatic system which plays a very crucial role of collecting the fluid that is left behind in various body tissues by the capillaries. This fluid is transported by this system to the region of the heart where it enters the heart through the *vena cava*.

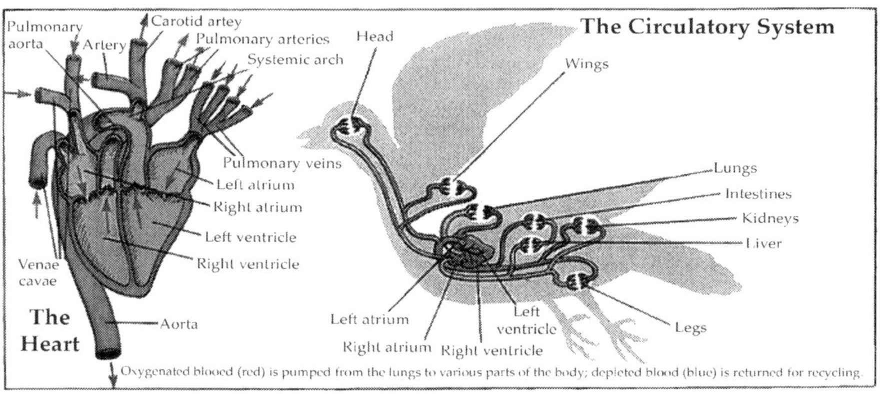

Figure 3.1: Avian respiratory system.

Source: *http://www.paulnoll.com/Oregon/Birds/Avian-Respiratory.html*

The Heart

The heart is the main pump of the blood in two distinctive directions- to the respiratory system (lungs) where carbon dioxide is removed and replaced with oxygen and pumping blood to the other body parts in order to deliver oxygen and other essential nutrients to the tissues and cells and collecting carbon dioxide and other waste products. From the heart, blood is pump through arteries referred to as aorta to the body and to lungs through pulmonary artery. From the body, the blood is subsequently pump back to heart through vena cava vein and pulmonary vein from the lungs.

The heart is located in the thoracic cavity of the birds, usually between the two liver lobes. The size of heart differs based on the size of the birds. It is also covered by a thin membrane referred to as pericardium. The poultry heart is subdivided into four key chambers (two atria and two ventricles) as it is seen in most mammals. Generally, the shape of the heart appears conical with a pointed end. Internally, there is a shallow groove which separates the small atria from relatively large ventricles.

There are four key organs linked to the circulatory systems and play vital roles in the system. These include the kidney, the lungs, the liver, and small intestines. The kidneys play the role of removing all the metabolic wastes and excess water from the blood stream. Just like in most mammalian circulatory system, most nutrients are absorbed in the small intestine and transported to the liver (for detoxification purposes) through hepatic portal vein. This special vein is different or separate from the hepatic system responsible for maintaining the liver. The lungs on the other hand help in removing the carbon dioxide and replacing it with oxygen. Blood from lungs is pump back to the heart through pulmonary vein.

Poultry Blood

Blood acts as the main vehicle for transportation in the body of birds and other animals. It is a very intricate compound consisting of suspended elements and intercellular fluid or plasma. Some of the suspended elements include the erythrocytes (red capsules), leukocytes (white capsules), and thrombocytes, which are compared to the platelets in various mammals.

The main function of poultry erythrocytes is to carry oxygen to the body tissue from the lungs and carbon dioxide to the lungs from various body tissues. Erythrocytes are produced in the red bone marrow. For every millilitre of blood in adult fowls, there are about 2.5 million erythrocytes. Leucocytes on the other hand help in fighting against various poultry infections. They are produced in the lymphoid tissues, spleen, and in some special cells in the bone marrow.

Most poultry leucocytes are classified in to five main groups; *heterophils* mainly for defencing the body against various bacterial infections, *eosinophils* are mainly used for defence against various parasitic infections and detoxification in the body of poultry, *basophils* they are mainly used for the degeneration of various eosinophils in the body of poultry, *lymphocytes* they are the most numerous of all the leukocytes and are responsible for detoxification of various toxic substances in poultry body. Lymphocytes are also responsible for releasing various enzymes that helps in nucleoplasm synthesis. They also release certain compounds known as gamma globulins believed to contain some antibody properties. *Monocytes* are the largest leukocytes and are responsible for the removal of various protozoans, bacteria, and other tissue debris.

Most poultry thrombocytes resemble mammalian blood platelets. They are the smallest blood cells. Similarly to the red blood cells, the thrombocytes, which are responsible for blood clotting, are also formed in the bone marrow.

There are other numerous factors and substances usually present in the poultry blood. These include antibodies, enzymes, metabolism waste products, hormones, and other products from the birds' organs and tissues.

Lymphatic System

This is a special poultry system that is linked to the main circulatory system. The term lymph results from the body fluid that is found between various interstitial spaces (i.e., spaces between various body tissues). The main function of this system is to drain any fluid that is left behind in various body tissues by the blood vessels. These fluids are then transported and drained into the *vena cava* of the main blood stream.

Spleen

The spleen is found on the right junction of the gizzard and glandular stomach of poultry. One of the key functions of this organ in the circulatory system is to screen all unsolicited particles from the blood. It is also responsible for the formation of various lymphocytes. It appears round in shape and reddish brown in colour. Structurally, it is a thin fibrous capsule shaped organ surrounded by some muscle fibres.

RESPIRATORY SYSTEM

This is one of the major systems in poultry and performs several functions including the removal of carbon dioxide, provision of oxygen, and thermoregulation (removal of excess heat and vocal communication). The respiratory system in poultry has similarities and differences when compared to that of the mammals. One major difference is that most birds do not have a diaphragm that usually helps in inflating and deflating the lungs. Some of the key sections of the poultry respiratory system include the nasal cavity, oropharynx, trachea, syrinx, bronchi, lungs, and air sacs.

Nasal Cavity

The nasal cavity has nares located at the base of the comb and on top of the mandible. It has a triangular shape and occupies the space between the eye margin and the nares within the beak. There is another structure called lacrimal sinuses which are located between the nasal cavity and the integument. This sinus opens through lateral wall into the cavity. Lateral wall has three major projections (conchae). These include the anterior projection which is a single-celled layer of squamous epithelium, the medial projection which is the ciliated epithelium columnar with hairs or cilia that help in trapping foreign materials, and the posterior projection of olfactory membrane responsible for sensing smell.

There are several mucosal glands located in the nasal cavity that also participate in trapping various foreign materials that attempt to invade the respiratory system. Nasal glands are also found in the nasal cavity and help in complementing the function of the kidneys through reducing the amount of body fluid electrolytes such as sodium chloride.

Oropharynx

The oropharynx is made of the birds' mouth and pharynx. The mouth roof (palate) has both soft and hard sections. The palate has a cleft opening called choanal. There is another opening behind the tongue of the birds called rima glottidis. This opening is connected to the cranial larynx. Larynx is a slit with arytenoid cartilages on the sides to offer structure support. Most of these cartilages resemble the mouth of pitcher. When the birds have some respiratory distress, the larynx can open up to 7–9 millimetres when gasping for air. Poultry does not have epiglottis, vocal cords, and thyroid cartilages, which are commonly found in most mammals.

Trachea

This forms a longitudinal section of respiratory system that helps in the movement of the inhaled air from the upper part of the system to the organs of respiratory system such as lungs and the air sacs. In adult poultry, the length of trachea can range from fifteen to eighteen centimetres. It is made up of cartilaginous rings which are arranged in a way to offer support to the trachea and preventing it from compressing or collapsing. These rings are also flexible to allow flexion and elongation of the trachea. Trachea has special structures called mucociliary epithelium which help in moving various foreign materials such as duct particles up and out of the system. on the tracheal lining there are several glands that secrete mucosa fluid that also help in trapping foreign materials as they try to enter in to the interior parts of the system.

Syrinx

The syrinx forms the vocal organ of the poultry. It is located near the ending of the trachea. It is made up of numerous pessulus, which are wedge-shaped cartilages. At the bottom of the syrinx, there are four to seven thin and flexible cartilages joined on one side to the pessulus and free at the other end. Sometimes these cartilages are referred to as syringeal ears or intermediate cartilages.

Trachea at the syrinx divides into two main bronchi which then connect to the lungs. The bronchi are made of C-shaped cartilaginous rings after the syrinx. Syrinx walls are made of vibrating membranes referred to as tympanic membranes. The sound made by the poultry is caused by these vibrating membranes. The sound produced by the birds is usually governed by the tension of the membrane which is controlled by various muscles along the respiratory system.

Bronchi

Bronchi are formed when the trachea divides at the syrinx. These are referred to as primary bronchi. There are the right bronchi and the left bronchi. One interesting part concerning these bronchi is that their combined cross sectional area is more than twice that one of the trachea caudal ending. Based on this allegation, airflow resistance and air pressures are significantly lower in bronchi than in trachea. Similarly, these bronchi are also made of cartilaginous rings until at the point they enter the lungs. The epithelium walls of the bronchi are also ciliated and have several mucous glands to help in trapping any foreign material that might have passed the other ciliated sections of respiratory system. Before entering the lungs, there is another division of the primary bronchi into four other smaller series of secondary bronchi. These four secondary bronchi are further subdivided again to form numerous parabronchi of the tertiary bronchi. Unlike the primary bronchi, the tertiary bronchi are not ciliated and have squamous epithelium. They appear as if they are arranged in layers as they enter the lungs.

The Lungs

Poultry lungs usually appear flattened and are found at the end of the coelom head. All the poultry and other avian animal do not have a diaphragm that separates the thoracic cavity from the peritoneal. Structurally, the size of avian lungs is about one tenth that of mammalian lungs, meaning that it is denser than the mammalian lungs. They are a pink in colour and are connected to the end of bronchi. The terminal structures of the lungs are referred to as the air sacs or vesicles. Compared to other animals, the lungs of the poultry are usually less elastic.

Air Sacs

These are balloon-like thin walled structures formed at the endings of the bronchi. Some of the air sacs in the lungs are usually connected to the long bones, thus forming the pneumatic bones, which renders the long bone lighter, an advantageous trait for most poultry to fly. Most poultry have nine distinct air sacs, including one clavicular air sac, two cervical air sacs, two cranial thoracic air sacs, two caudal thoracic air sacs, and two abdominal air sacs.

Most of the clavicular air sacs are located at the coelom cranial endings. Each of these air sacs has two lateral chambers and one median chamber. The median chamber is linked to the coracoid, sternum, and the ribs of the birds. On the other hand the lateral chamber is linked to humerus of the birds. The cervical air sacs are found at the coelom front end in the dorsal areas of the birds. Most of them are linked to the thoracic vertebrae, cervical vertebrae, and the first two ribs of the birds. Cranial thoracic air sacs on the other hand are found in the thoracic cavity along the dorsolateral region of the birds. Caudal thoracic air sacs are densely located behind the thoracic cavity. Lastly, the abdominal air sacs are relatively larger than other air sacs and are found along the cloaca region on the lungs rear sides.

Because most of the birds' air sacs are transparent and thinned walled, it is quite difficult to observe them with the naked eye. Apart from the abdominal air sacs which are linked directly to the primary bronchi, all the other air sacs are connected to the secondary bronchi and tertiary bronchi of the birds. During inspiration and expiration of the birds, the air sacs are responsible for supplying huge volume of air in and out of the lungs. Most of these airs sacs are also connected to various long bones of the birds where they supply the oxygen used for powering the birds during flight.

REPRODUCTIVE SYSTEM

Most poultry birds have a heterosexual reproductive system, indicating that both sexes are required for the input towards the genetic structure of their offspring. The male birds offer the half of this structure through producing sperms in their testes, while the other half is offered by the female birds through the ovum produced in the ovary. The ovum in poultry is commonly referred to as the germ disc, blastoderm, or blastodisc. Once the sperms are released in the testes, they are carried by the semen. On the other hand, the ovums are carried through egg yolk once they are released in the ovary.

Male Reproductive System

The reproductive system of male poultry birds usually comprises of two testes. Each testis has a deferent duct that run from the testes and ends in the cloaca of the bird. As opposed to other animals, male poultry do not have penis. The testes are bean-shaped and are located in front of the kidney next to the backbone of the male bird. The size of the testes is usually inconstant and enlarges when the male birds are actively involved in mating. The right testis is usually smaller than the left. Internally, each testis has a flattened portion that is thought to resemble to the mammalian epididymis.

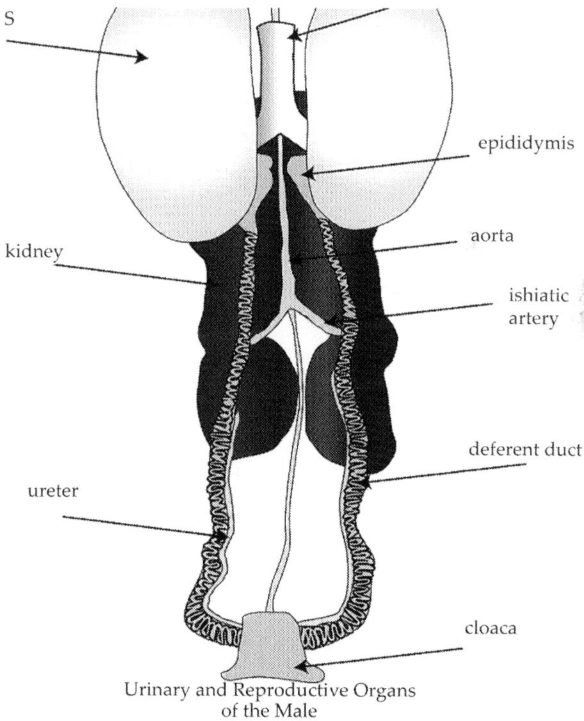

Figure 3.2: Reproductive organ of a male poultry bird.

Source: *http://www.poultryhub.org/physiology/body-systems/reproductive-system/*

Deferent duct: This is a tube that helps in transportation of sperms from the males testes (where they are formed) to the males' cloaca (where they enter into the female oviduct during mating). In the cloaca, deferent duct enters a pimple-like organ resembling the mammalian penis. This structure in ducks is usually big forming a penis-like organ. Deferent duct is usually contracted as it leaves the testes but broaden as it enters the male cloaca.

Testes and sperm: Both testes have twisted ducts known as seminiferous tubules. These tubules facilitate formation and transformation of sperm through a process referred to as meiosis. The sperm is responsible for half of the required chromosomes to form an embryo. The other half is produced by the female bird. There are about three to five million of sperm in one cubic millimetre of semen produced by male birds. The testes are also responsible for the production of special hormones known as androgens which stimulate development of various secondary sex traits of males such as mating behaviours, combs and wattle growth, and other male conditions.

Female Reproductive System

The main parts of female reproductive system in poultry are the ovary and oviduct. Only the left ovary is functional. The ovary of egg laying birds is located in the abdominal cavity just in front of the kidneys. It is usually attached firmly to the abdominal cavity. The ovary is usually well served with highly dense network of blood vessel to facilitate enough supply of nutrient for the growth and development of egg yolk.

Ovary: There is a yellowish, round-shaped mass in the ovary referred to as follicle. Each of these follicles contain one ovum or the egg yolk. There are numerous follicles, but only a few usually reach maturity to produce eggs. The ovary of the female birds is active only during the laying season. For the follicles to develop to maturity, they have to undergo various stages of development.

- Primary follicle stage- involve all those follicles that have not yet started growing
- Growing or developing follicles
- Maturing follicles-involves all those follicles that are ready or nearing to be released

- Discharged follicles-involves those follicles that have released the yolk
- Atretic follicles-involve all the follicles that have released the yolk some time ago

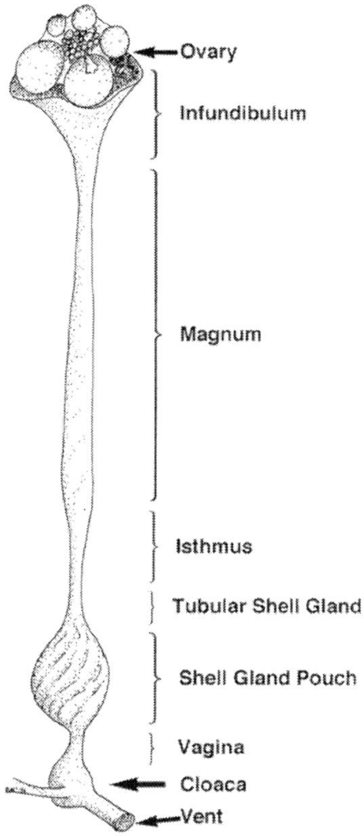

Figure 3.3: Reproductive organ of a female poultry bird

Source: *http://www.poultryhub.org/physiology/body-systems/reproductive-system/*

Yolk: It takes around 10 days for an egg yolk to develop fully into the normal size seen in laid eggs. Egg yolks are produced and developed in follicles where necessary nutrients are supplied for their growth and development. When they are fully mature, the follicles usually split on the stigma to release the yolk into the oviduct. If the follicle split on surfaces other than on the stigma, most blood vessels usually burst resulting in blood spots on the eggs.

Oviduct: The main function of the oviduct is to generate the shell, membrane, and albumen for the structural formation and development of the egg. Structurally, this is a long duct with numerous blood vessels. There are also several glands found on the walls of the oviduct. These glands are responsible for the production of the shell, shell membrane, and albumen. Most non-laying poultry birds have a smaller and shorter oviduct. However, once the birds reach reproductive stage, this oviduct usually grow to about seventy to ninety centimetres in length. The oviduct is made up of five different sections and each function differently. These include *infundibulum* (or the funnel) where the egg yolk is collected once it is released by the follicle. It is also the point where fertilization of the ovum takes place. *Magnum* is responsible for the secretion of albumen. *Isthmus* is responsible for the secretion of both shell membranes and some albumen. *Shell gland* is responsible for secretion of shell and some albumen. The *vagina* is responsible for pigmentation of egg shell and secretion of outer cuticle of an egg.

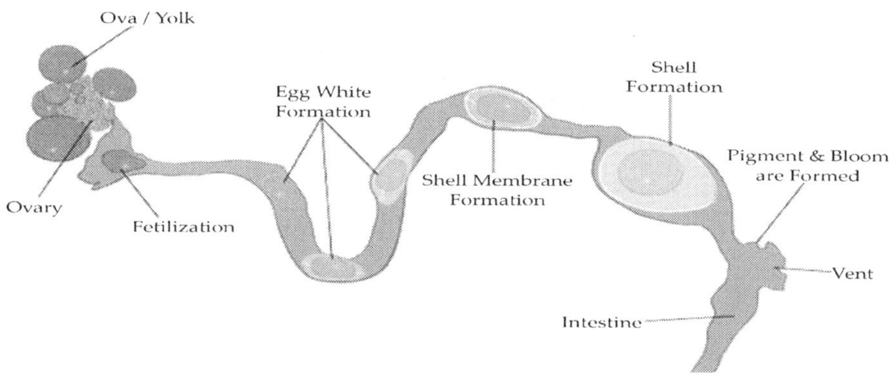

Figure 3.4: Different processes egg formation with the oviduct.

Source: *http://www.cappersfarmer.com/farm-and-home/poultry/mommy-where-do-eggs-come-from*

Apart from formation and production of eggs, the reproductive system of female poultry is also responsible for the production of various hormones that help in regulating and controlling various functions of the body. Some of the key hormones include oestrogens, androgens, and progesterone. Androgen hormone aids in formation of albumen, the growth of the combs, and other conditions. Oestrogen on the other hand is responsible for the development of oviduct, albumen and egg formation, plumage growth in females, nesting and mating behaviours among other functions. Progesterone in combination with androgen is responsible for albumen formation and stimulating pituitary gland in order to release luteinising hormone.

EMBRYOLOGY OF THE CHICKEN

Chicken embryology is the growth and development of young birds inside the egg. It is important for poultry keepers to try and differentiate abnormal embryos from normal embryo formation in the eggs, and what might result to mortality of embryo during incubation period.

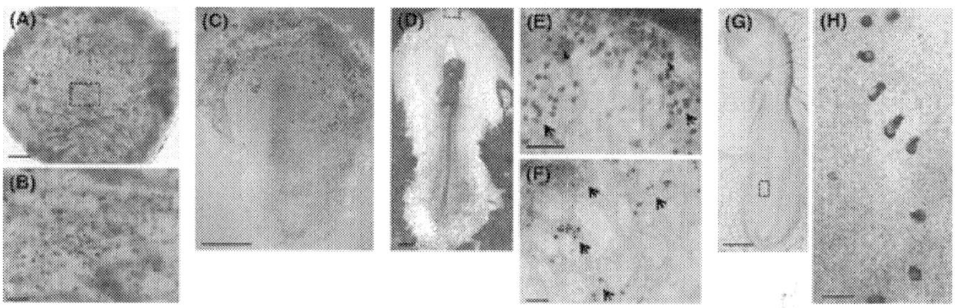

Figure 3.5: An overview of the early embryology in chicken from 18 to 72 hours after fertilization

Source: *http://journals.plos.org/plosone/article?id=10.1371/journal.pone.0092672*

Fertilization and Early Embryo Development

Fertilisation of egg and early embryo development usually starts when a male bird mates with a female bird. The germinal disc in the ovum or egg yolk of a female bird is fertilised by a sperm in the infundibulum about fifteen minutes after it is released by the follicles. After fertilization has taken place, new embryo starts to form through cell division about five hours later. This is a continuous process that takes place even after the egg is laid by the female birds. Under favourable incubation condition, fertilized eggs usually take roughly 3 weeks for the chicks to develop fully and hatch.

When the male sperm cell (one half of the necessary chromosomes) fertilises the ovum or the female egg cell (the other half of the necessary chromosomes) usually leads to formation of a zygote. This is a single-celled creature comprising of all chromosomes. Roughly five hours once fertilization has taken place, the formed zygote moves to the isthmus region, which is the place that the embryo usually begins to develop through cell division process. The embryo comprises about eight cells by the time it leaves the isthmus. It is then moved to the uterus where it spends about fours where it grows to about 256 cells.

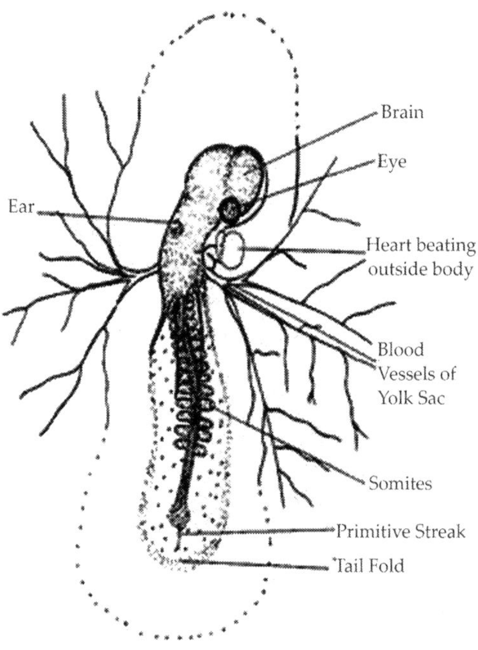

Figure 3.6: Different parts of an avian embryo in the early stages of development.

Source: *http://www.poultryhub.org/physiology/body-systems/embryology-of-the-chicken/*

During the initial development of an embryo, only one layer of the cell in the yolk is formed, but sometimes two layers are formed after a continuous process of cell division. These two layers are referred to as the endoderm (underneath layer) and ectoderm (uppermost layer). They are both formed after the blastoderm central cell separates itself from the yolk, thus forming a cavity. This cavity enables the development of the subsequent embryo. Once the ectoderm and endoderm are formed, another third layer commonly referred to as mesoderm (middle layer) is usually formed. These three layers form the foundation of various organs and tissues of the bird.

The ectoderm is responsible for the formation of birds' skin, the feathers, nervous system, claws, and beak. The endoderm on the other hand is responsible for the formation of various systems in the body of the birds. These include the secretory system, digestive system, and the respiratory system. Mesoderm is responsible for the formation of excretory system, various organs of reproductive system, muscles, circulatory system, and the skeleton of the birds. By the time an egg is laid, the embryo usually consists of all cells differentiated into various body systems, organs, and tissues.

Extra-Embryonic Membranes

Since poultry embryos usually do not have structural connection to the mother bird, all the necessary embryonic nutrients apart from oxygen are confined in the eggs. During the early development of the embryo, there are special external membranes formed by the embryo in order to access its nutrients from the egg and execute all significant structural functions. During the embryo development, there are four key membranes that are formed within the egg. One of these membranes is the *yolk sac*, which is mainly formed to cover or envelop the yolk. It also secretes a very crucial enzyme that helps in transforming various material of the yolk into consumable material for the developing embryo. All the remaining materials within the yolk sac are collected into the embryos abdomen to be used by the newly hatched chick for the first 2 to 3 days before the chicks are able to distinguish what to eat or drink and where to actually find some of its feeds.

Apart from the yolk sac, there is another membrane called the *amnion membrane*. This membrane provides the medium in which the embryo floats in. In other words, the amnion membrane acts as a shock-absorber where the delicate embryo grows and develops without being harmed by various environmental conditions. The third membrane is called *allantois membrane*, and it participates in developing an extensive network of circulatory system that is linked with the embryo. The membrane is propelled by the heart of the embryo. Apart from circulatory system, allantois membrane is responsible for all excretory and digestive functions that take place in the embryonic system. The fourth and last membrane is known as *chorion membrane*. This membrane is mainly responsible for connecting the allantois membrane to the inner shell membrane in order to conduct all its functions.

Daily Embryonic Development

It is important as a poultry breeding farmer to familiarize yourself with various stages of embryo development from day one until fully developed chicks are hatched. This helps in understanding or monitoring the hatchability in poultry. It also helps in determining the stage of embryonic development with highest mortality. There are different stages of embryo development during incubation period from the time fertilization of the ovum takes place in the ovary until the day the egg hatches in new chicks. These stages are illustrated in the following table.

Table 3.1: Different stages of embryo development during incubation period from the time fertilization of the ovum takes place in the ovary until the day the egg hatches in new chicks

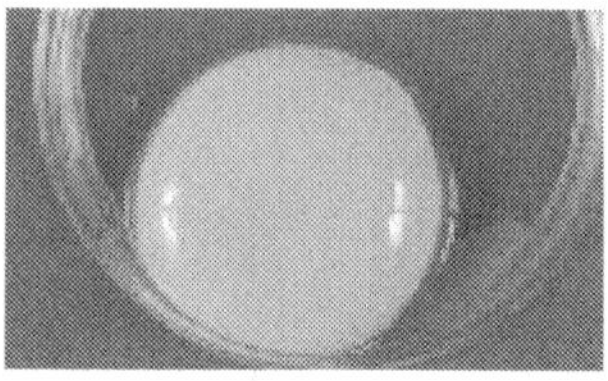	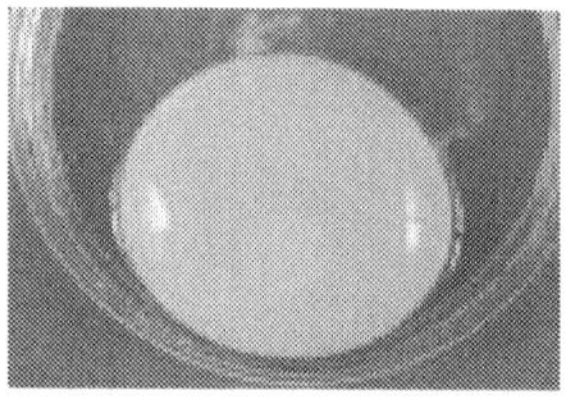	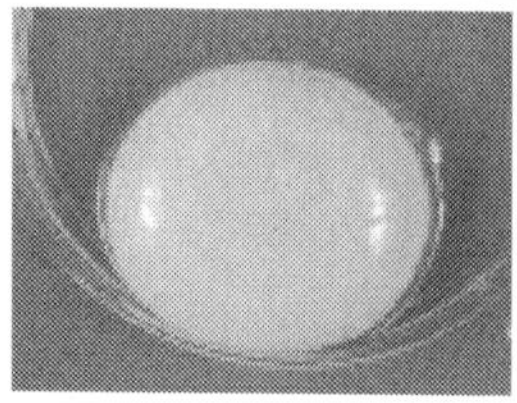
Infertile stage No any form of development	**Day 1:** Tissues starts to develop within the yolk	**Day 2:** Tissue development starts to be more visible plus appearance of blood vessels

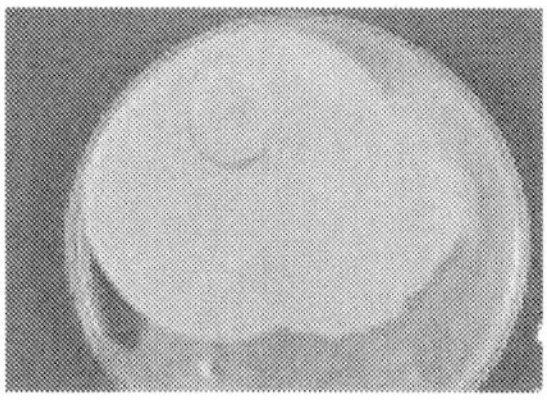 **Day 3:** Some hearts beats are visible plus more visible blood vessels	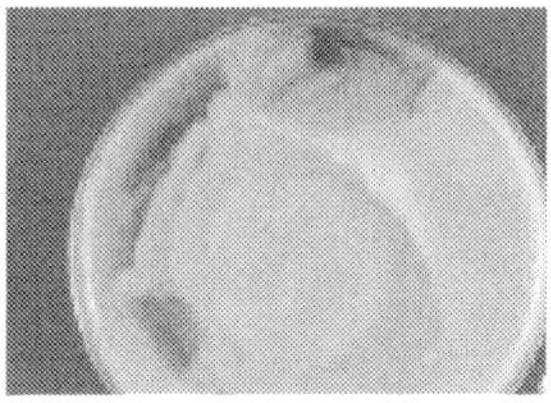 **Day 4:** Some pigmented eyes starts to be visible	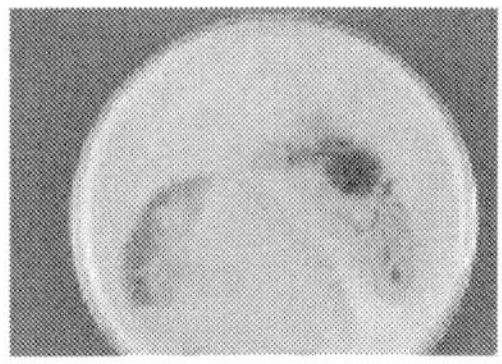 **Day 5:** There is some appearance of elbows and knees
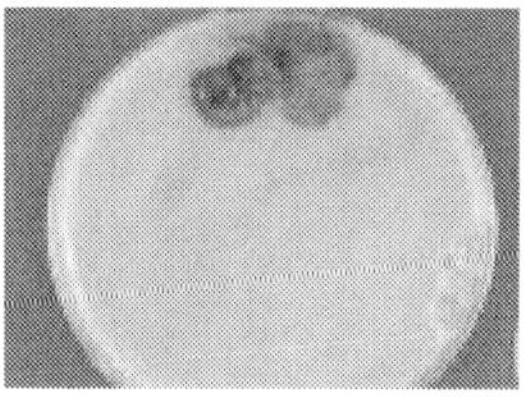 **Day 6:** The beak start to develop plus some voluntary movements are also seen	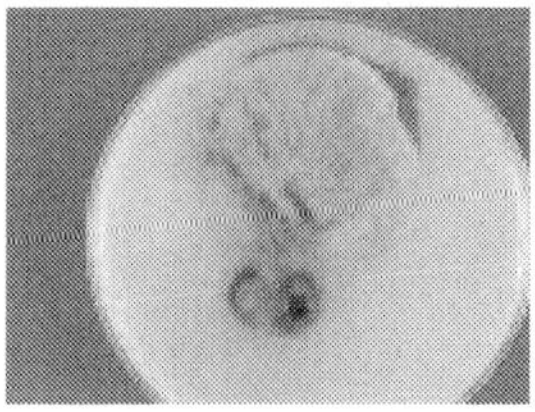 **Day 7:** The comb starts to develop plus appearance of the egg teeth	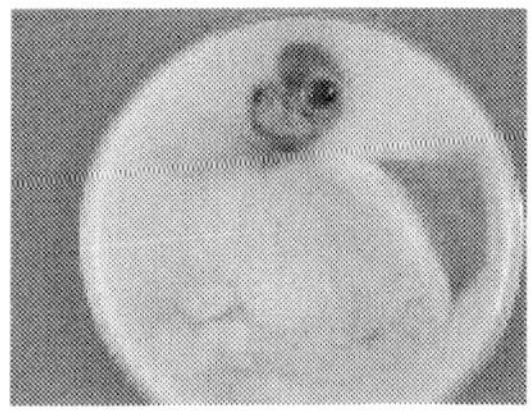 **Day 8:** Some feather tracts start to appear plus equal lower and upper beak becoming more visible
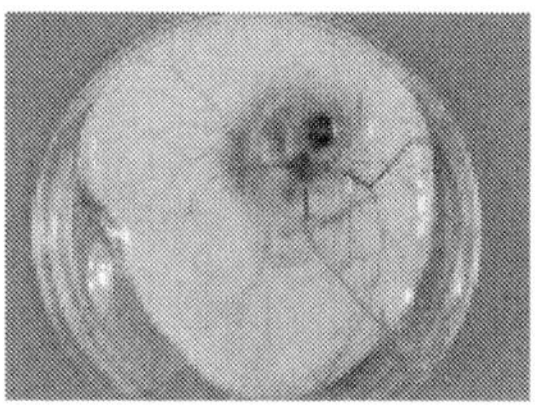 **Day 9:** The embryo now start looking like a real bird plus mouth opening start to be visible	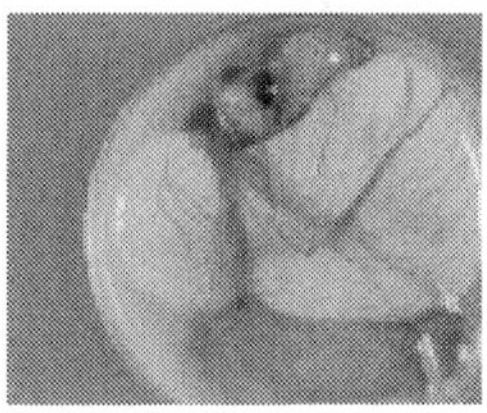 **Day 10:** The egg tooth is now prominent plus toe nails start to develop	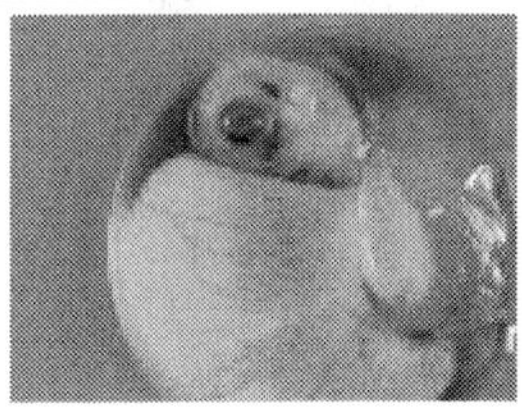 **Day 11:** Serrated comb become visible and the tail feathers start to appear

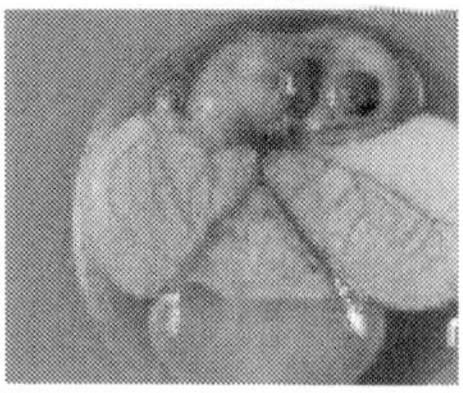	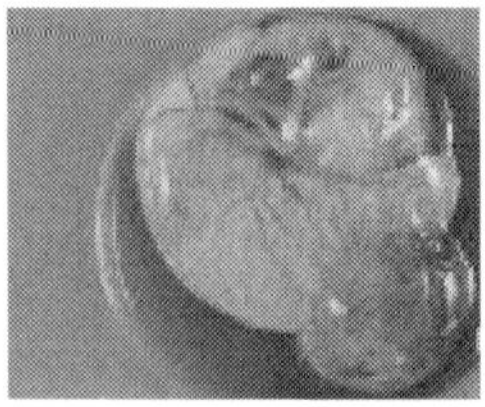	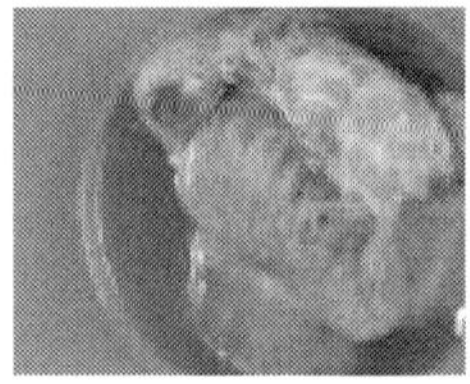
Day 12: Toes are fully developed and first few feathers are visible	**Day 13:** Scales start to appear plus partial body covering with feathers	**Day14:** The head of the embryo moves towards the large end of the egg
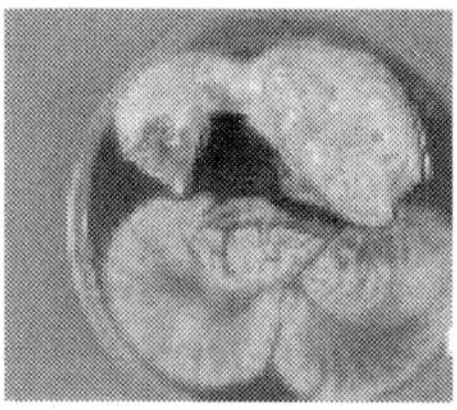	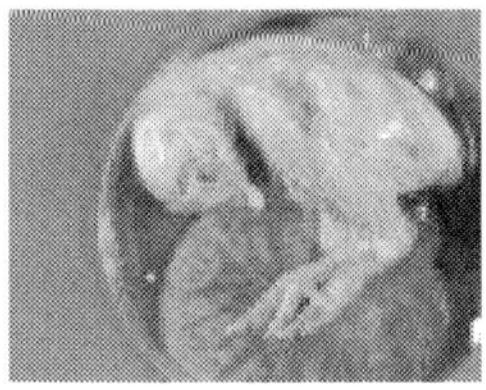	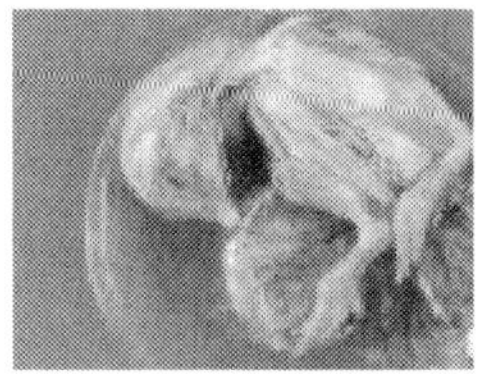
Day 15: The gut is now drawn into the abdominal cavity of the embryo	**Day 16:** The body is now completely covered with feathers plus the albumen nearly diminish	**Day 17:** The head is between the legs plus all the amniotic fluid reduced
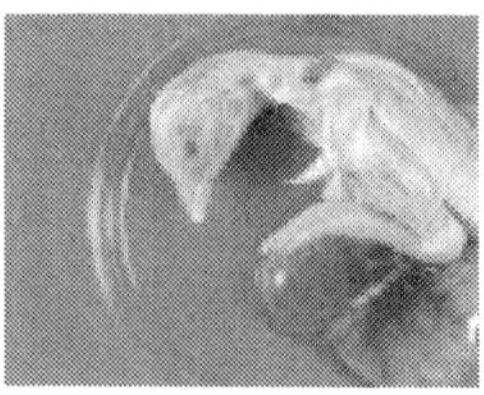	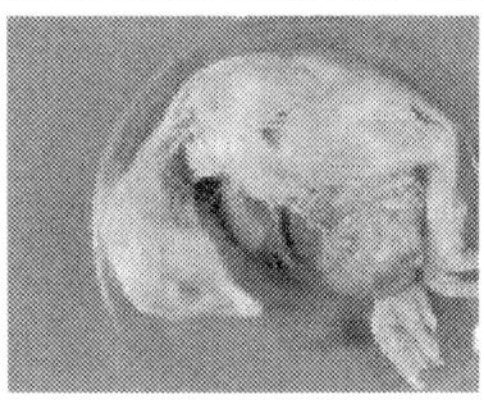	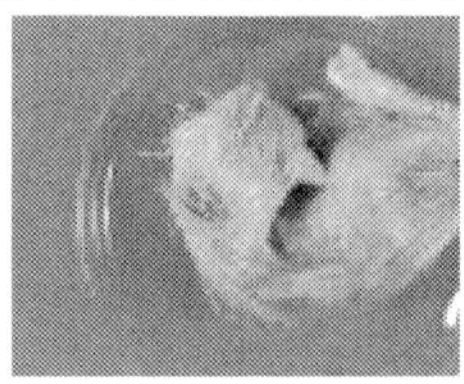
Day 18: Growth of the embryo almost complete, the yolk sac still on the outside of the embryo and the head is under the right wing	**Day 19:** The yolk sac is now drawn into the abdominal cavity, all amniotic fluid is finished, and the embryo now occupying most of the space in the egg	**Day 20:** The yolk is completely drawn into the body of the embryo, the embryo is now breathing in air cell and the internal and external pip

Source: *https://www.pinterest.com/pin/65724475789373063/*

Once the all the stages are complete, the embryo assume the right hatching position, where by the front part of the embryo's body is drawn towards the large side of the egg, the head of the embryo is position underneath the right wing, and the legs are positioned underneath the head.

NERVOUS SYSTEMS & IMPORTANT SENSORY ORGANS

The nervous system and various sensory organs in the body of a bird play a very crucial role in the daily functioning of the birds. In other words, the nervous system is responsible for the control and integrating different body functions of the birds, while the sensory organs are responsible for detecting various stimuli in the environment of the bird thus helping them to react accordingly. These functions are either voluntary (actively) or involuntary (automatically) regulated.

Voluntary controlled functions occur when the birds response to various stimulus or activities. Based on these stimuli, they have a choice to make an action such as moving particular parts of the body. The nervous system responsible for the voluntary responses is referred to as somatic nerves.

Involuntary controlled functions occur when a bird have no choice based on the occurrence of a situation. In other words, involuntary responses occur when the birds lack conscious control over a certain event. Some of the functions in the body of a bird that respond to involuntary control include digestion of foods, body circulation, heart beat regulation, and respiration activities. This means that poultry birds are not able to control the digestive functions in their bodies, heart beating, or even their breath. All these crucial functions in the body of the birds are only controllable using the automatic nervous system.

The Brain

Similarly to other animals, avian brain is located in the head and is usually well governed by cranium bones in the head. The brain of bird is subdivided in to a number of sections. It acts as the main or central control for almost all functions and activities in the body of a bird. Some of the major parts of the avian brain include: the *forebrain* which is made of the olfactory lobes and the cerebral hemispheres. Both pituitary gland and the hypothalamus glands are found below the forebrain. The *midbrain* which is made of the optic lobes is the other part of avian brain. The *hindbrain* is made of medulla oblongata and the cerebellum.

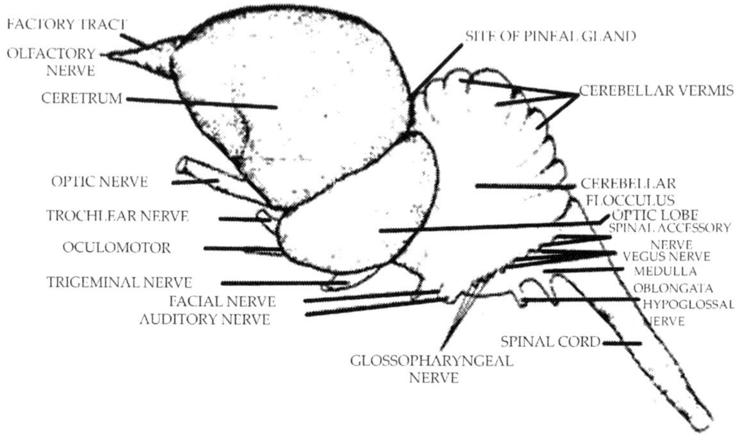

Figure 3.7: Different parts of an avian brain.

Source: *http://www.poultryhub.org/physiology/body-systems/nervous-systems-important-sensory-organs/*

Olfactory lobes which are located in forebrain act as the main receptor region for the olfactory nerves responsible for smell detection. On the other hand optic lobes are the receptor regions for various optic nerves responsible for poultry eye sight. Based on the total size of the brain, poultry have the biggest optic lobes compared to other species. This suggests that eye sight in poultry is a very crucial role for their day-to-day activities.

The pituitary gland, which is directly related to the hypothalamus gland, plays a significant role in the endocrine system of the birds. This gland is commonly referred to as "master" gland because it is responsible for regulating all the functions of different systems, organs, and different tissues of the birds.

Spinal Cord

This is a cord made up of a network of nerve tissues extending from the brains medulla oblongata to the vertebral column of the birds. Both the brain and spinal cord form the central nervous system of the birds. Just like the brain, poultry spinal cord is also highly protected by the spinal fluid and a special sheath. It is then enclosed within vertebral column. Spinal cord has a densely compact network of verves that control various functions of body systems, organs, and body tissues. From the spinal cord, these nerve leave through various opening found in between the joints of various vertebrates.

If some parts of spinal cord were infected or broken, the linkage between body systems it control and the brain will also be broken hence losing control. For instance, if the lower back of poultry spinal cord is broken, the birds may end up losing control on the legs and other parts below the broken part (paralysis).

Poultry Sensory Organs

Various sensory organs of a bird usually receive different stimuli from the surroundings of the birds. There are different sensory organs in the body of birds; these include the eyes responsible for sight, ears responsible for hearing and balancing, taste buds responsible for tasting, and olfactory organs for smelling.

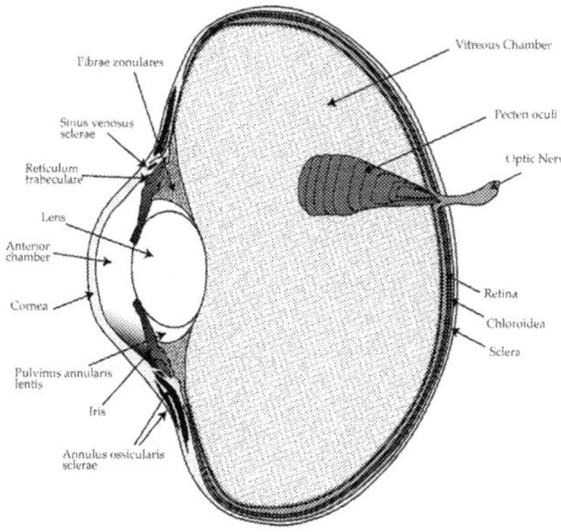

Figure 3.8: A cross-sectional structure of an avian eye

Source: *http://www.poultryhub.org/physiology/body-systems/nervous-systems-important-sensory-organs/*

Compared to the size of the birds, poultry eyes are usually large. Each eye is located on both side of the head, and about 25 degrees straight to the front of the bird's face. Each has two key eyelids and another membrane referred to as nictitating membrane situated in the front corner of the eye. When the eyes of the birds are closed, this membranous eyelid is usually seen partially. The eyes of the birds are similar to those of other mammals but they have more thresholds to the intensity of light. This means that birds are able to see clearly over a long distance than the mammals. This actually enables the birds to be more active in the day time (diurnal) rather than during the night (nocturnal).

Ear

The ears of the birds are located behind the eyes. They are much smaller as compared to those of other mammals. The birds' ears usually do not have ear flap, pinna, or ear lobes.

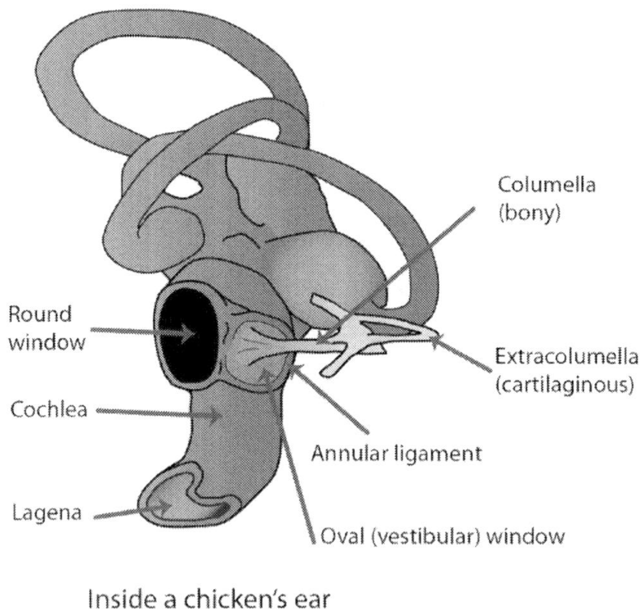

Figure 3.9: Illustrating the inside section of a chicken's ear.

Source: *http://www.poultryhub.org/physiology/body-systems/nervous-systems-important-sensory-organs/*

Poultry ears are divided into three main segments. These include the *outer ear*, which is made up of ear canal. There are also some feather tufts on the outer ear that protect the ear from various foreign and potentially harmful objects. The *middle ear* is separated from the outer ear by another membranous canal called tympanic membrane. The *inner ear* is made of a special nerve called cochlea which helps in receiving all the sound waves for transmission to the brain. The inner ear also has another special part called semi-circular duct, which helps in maintaining the balance of the birds. There is another duct that connects the roof of mouth to the middle ear. This duct is responsible for regulating the environmental air pressure to that of inner ear, thus preventing injuries caused on the tympanic membrane.

Olfactory Organs

These are poultry organs responsible for smelling. These organs in poultry are considered as not very essential because smell is of secondary significance compared to the other poultry senses. Although this is this case, poultry selection to feeds and other daily behaviours does not change.

Taste Buds

Taste buds in poultry are situated underneath their tongue and on the floor of their mouths. They have special glossopharyngeal nerve endings. These taste buds are used to recognise taste of certain chemicals and transfer the information to responsible brains receptors. These buds are usually more sensitive or active on the taste of material in liquid form than those in dry form. Poultry also have another special feature that enhances their sensitivity to the temperature of drinking water. This means that most poultry are likely to decline drinking water that is more than 33 °C.

EXCRETORY SYSTEM

All domesticated fowls consist of two kidneys in their excretory system. Each kidney has a ureter which transports the urine secreted by the kidneys outside the body through the cloaca. In case the kidneys are infected by any disease or they are damaged, they are all rendered inefficient and they cannot function as they are expected. This condition causes sudden death to the affected birds because the birds are usually weakened and cannot survive as a consequence. The following are some of the functions of the excretory system in the domestic fowl: maintaining the balance of the electrolyte in poultry body; it removes excretory waste from the body, this waste includes nitrogenous products with exception of the carbon IV oxide gas; it also maintains the balance of the water in the body system of poultry.

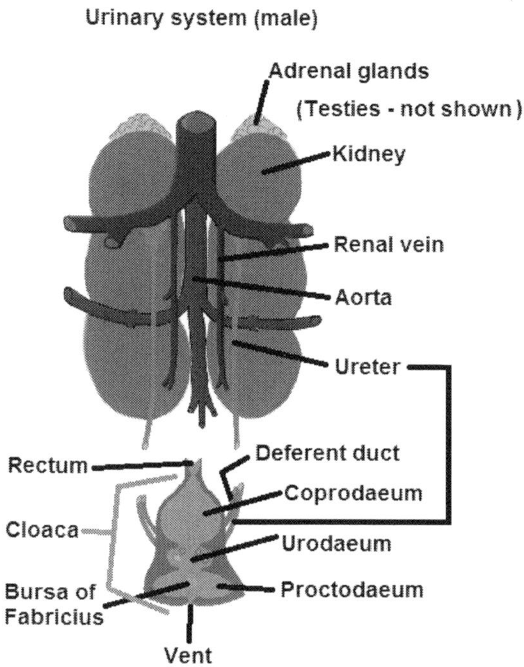

Figure 3.10: Urinary system of a male bird

Source: *http://www.poultryhub.org/physiology/body-systems/excretory-system/*

The Kidneys

The birds' kidneys are usually red-brown in colour and they usually possess three lobes in general. They located exactly at the back of the lungs on each side of vertebral column in which they have an association with it. The ureter, which appears straight and narrow tube, it leaves the central border of each of the kidney and it further extends to the cloaca, which lies next to the deferent duct of the male or the fallopian tube of the female. The ureter offers connection from each lobe of the kidney to other funnel-shaped structures found in the kidney. When inspection is done on the kidney, the renal cortex and the renal medulla can be seen clearly in the kidney. The microscopic examination reveals that the kidneys possess a huge number of renal tubules, or in other words nephrons, which are all separated into cortical and medullary parts. Every bird possesses two types of nephrons. In the renal medulla, there are nephrons with a loop of Henle for the purpose of concentrating urine. These nephrons resemble those ones of a mammal. Additionally, there are other nephron types in the renal cortex. These nephrons usually do not have loop and they resemble those ones of reptiles. The renal tubules serve the function of extracting the components of the urine away from the blood which flows via the kidney.

The following are the composition of the nephron:

- Renal corpuscle: This is characterised by composition of blood capillaries which have close network and they are all enclosed inside a capsule, commonly referred to as Bowman`s capsule. It is in this renal corpuscle where the inevitable close connection with the blood takes place.
- Proximal convoluted segment: which is spiral like in shape and it is from the bowman's capsule.
- Loop Henle: which is straight and it is only found in the nephrons, in the renal medulla which resembles those of a mammal.
- Distal convoluted segment.
- Collecting tubule: which serves the purpose of transporting urine inside the ureter for excretion out of the body.

Urine

The domesticated fowls do not have a bladder, which is found in many amphibians and mammals. From the ureter, the urine departs and it is deposited in the cloaca, whereby it is later transported by the reverse peristalsis towards the large intestine. The large intestine allows the reabsorption of the excess water before the urine is excreted from the body. Reabsorption of excess water back into the large intestine is very useful to the bird as it helps to outset the limited stability of the birds to pay much of their attention to the urine as it is in the mammals. The urine of birds is usually in form of thick paste, which contains very little water but contains a high concentration of uric acid from nitrogen metabolism. This kind of urine is passed out as a paste and usually appears as a whitish or creamy-like cap on some of the faecal stools of the birds.

Figure 3.11: A whitish uric-acid the faecal stool of a poultry bird.

Source: *http://www.fresheggsdaily.com/2016/03/all-about-chicken-poop-brown-black-and.html*

Kidney dysfunctioning or high levels of proteins deposits into the kidneys leads to huge quantities of uric acid inside the blood, which may cause dysfunctioning of the excretory system. This condition causes the kidney tubules to swell because of the urate deposits. When this occurs, it is easily detected through observing the white lines that appear on the kidney's surface. This urates deposits cause damage to the kidney cells thereby causing nephritis. Once the uric acid is highly concentrated in the blood, it may lead to filtration via the capillary walls into visceral gouts. When this happens, white deposit can be seen on the surface of most visceral organs. These effects may also occur due to an infection in the bronchitis.

SECTION-III
POULTRY BREEDING

CHAPTER 4

IMPORTANT ASPECTS OF POULTRY BREEDING

INTRODUCTION

Based on the productivity capacity such as the quality and quantity of the meat and the size of the eggs, modern poultry are much superior compared to their ancestors. This has been made possible through various technological advancements in breeding techniques and other environmental factors. Poultry breeding can be described as "the scientific practise which aims at genetic improvement of the birds through progressive generations by virtue of planned production." In order to come up with the best traits in terms of desirable quality of the breed, poultry keepers must familiarize themselves with the various capabilities and qualities of different poultry breeds.

Possibility of getting high and quality breeds at village level is practically attainable if the right methods and techniques are employed amicably. Poultry farmers with little knowledge on the way forward should seek for information from experience and knowledgeable poultry breeders. However there are several aspects of poultry breeding that every poultry keeper should be aware of. These aspects are discussed in the simplest and systematic way in this chapter.

PRINCIPLES OF POULTRY BREEDING

There are several fundamental principles that every farmer should follow when rearing or breeding poultry in the farm. Every farmer should be aware of the importance poultry breeding and various standards in which birds in the farm should be bred. Some of the prerequisites of poultry breeding may include improving the weight and size of the eggs, improving eggs production, improving the quality of poultry meat, or a combination of all these poultry traits. For instance poorly bred female poultry are usually voracious feeders, but they do not lay congruently more eggs because they are not bred for the production of eggs. Efficiency and effectiveness in egg production based on feed conversion is a trait that is inherited and can only be enhanced through careful poultry selection and breeding.

When breeding poultry, it should be done from adult females and males that conform very closely to the desired standard. When selecting breeding birds, all those that fail to conform or possess the required poultry standards should be culled. It is also very crucial to always use pure breeds when selecting and mating poultry breeds. In other words, successful poultry selection and breeding should be done carefully and continuously from the hatching to maturity of the birds. It is important to remember environment influence the quality of poultry breeds attained. Therefore, creation of all favourable conditions in poultry is highly advisable. These include proper poultry feeding, housing, hygiene, and general management. For the farmers with more advanced technical expertise in poultry keeping, they can practise pedigree breeding, this ensures poultry selection and mating practices are efficiently achieved.

TYPES OF POULTRY MATING

Mating can be described as the practice of connecting a female bird with a male bird in order for the females to lay or produce eggs that are fertilized for the purposes of hatching and reproduction. There are several factors that determine the number of female birds that are allowed to be served by a single male. These include the weight of the male, the breed, age, season, physical condition, and virility of the male bird. For instance, more female birds may be served by a male in the White Leghorns (which are light breeds) than in the case of Rhode Island Reds (which are usually heavier breeds). Likewise, a young male (cockerel) can be able to serve more female hens (pullets) than in the case of an old cock. The number of females in mating pen can be increased during the spring seasons than during the winter. Because fertility and birds' activeness are usually very poor during summer, it is advisable to suspend poultry mating during this.

There are five different kinds of poultry mating that every farmer can use when breeding birds in the farm. These include artificial insemination, pen mating, stud mating, flock mating, and alternating males. Due to lack of expertise and infrastructure at the village level, artificial insemination is usually not the best proposal for poultry farmers.

Pen Mating

Involves putting about ten female hens in a breeding pen and only one male cock is allowed to live and mate with them generously. All the laid eggs after a week of mating are usually fertile and ready to be used for hatching purposes.

Stud Mating

This involves rearing all the females and males in separate pens. The females are then allowed to enter into the males' pen one after the other in a regular intervals. They are then removed after successful mating.

Alternate Males Mating

Two different male birds are used for mating. However, one male is permitted to mate with the females while the other male is restricted in a coop for the first day. The second male is then allowed to mate with the females in the following day. It is actually difficult to determine offspring paternity in this method.

Flock Mating

This is commonly practiced by many poultry farmers at the village level, and involves keeping a huge female flock with several males in a ratio of one male for every 10 females. In this kind of mating, some males may develop superiority through fighting others. The "superior" males become the antagonist and prevent other males from mating with the females. Therefore, the fertility of the birds may be negatively affected.

It also becomes practically impossible to trace which male is responsible for fertilization of the laid eggs, thus making pedigree poultry breeding impossible.

SYSTEMS OF POULTRY BREEDING

There are several modern systems used in improving the quality of poultry. These systems include inbreeding, outcrossing, line breeding, cross breeding, and grading-up system. Inbreeding involves close blood relations mating. A good example of inbreeding involves offspring to parents mating or sister and brother mating. If this kind of breeding system is used for a long period of time, it can lead to several negative consequences to the bred poultry. These consequences include poor hatchability, retarded growth rate, reduction in egg production, and short poultry lives. However, if parent birds have good traits, these traits may be inherited by the offspring.

Inbreeding and line breeding are similar to some extent, but for the case of line breeding, it involves poultry breeding that are less closely related. Some of the best types of line breeding include mating grandsires and granddaughters. In this system of breeding it is possible to enhance the quality of the birds in terms of the eggs sizes, more egg production, and enhanced body weight among other traits.

Cross- breeding system involves using different birds' breeds for mating. Based on this type of breeding, the resulting offspring is referred to as cross-breed. There are several advantages related to poultry cross-breeding system. Some of these advantages include good hatchability, reduced mortality, and faster weight gain among other traits. Cross-breeding always results in hybrid vigour, which has been adversely employed in meat-type (broiler) poultry production.

Out-crossing involves the use of same birds' varieties but different strains for mating. The aim of this system of breeding is to retain good characteristics of certain breeds in one family line and apprehending other good traits from a different family line or strain. For instance, a certain white leghorn strain may have traits of high egg production but small eggs size. Another white leghorn strain may have traits of producing larger eggs but reduced production. Through cross breeding these two different poultry strains, it is likely that the egg size will increase without the change in egg production.

Lastly, grading up involves the use of a purely bred male to mate a female of unknown breed in order to enhance the progenies traits of the female bird. If this system is to be used in poultry breeding, farmers are always advisable to change the purely bred male bird every year in order to prevent inbreeding. Additionally, other male breeds should not be allowed to mate with the females to prevent diluting the effects. Changing the male every year to mate with pullets of successive generation helps in improving the quality of chicks in every generation. In other words, chicks in the sixth generation are likely to possess the same traits of pure bred male used during the grading process. This is one of the best methods used by village poultry farmers to improve the qualities of their birds.

POULTRY SELECTION AND IMPROVEMENT

Selection means obtaining birds that possess desirable traits such as poultry meat production, egg production, good breeding traits, and exhibition traits. The success of any poultry farming investment is highly influenced by the quality of the selected and bred birds. This can be compared to poor crop yield as a result of the farmer selecting and using poor seed when planting. Similarly, a poultry farmer who uses poor males and females for the purpose of breeding will definitely incur huge losses even if proper management, good feeding, and disease control are practiced.

When selecting the type of birds to breed, there are several fundamental elements that every farmer is encouraged to adopt in order to set a good foundation for the stock.

- All hereditary characteristics of the breed used must be known
- Maturity age of the breed used must be know too, this helps in excluding all slow maturing breeds
- Feed conversion efficiency and egg production should be high for the breed (poultry feed consumption in every egg produced should be less)
- Growth rate and production of the breeding birds should be constant

- Good strain performance based on prevailing conditions
- Determining whether continuous stock replacement would be available

Selection for Egg Production

When selecting birds for egg production, there are certain considerations that every poultry breeder should make. Some of the traits to consider for good egg laying birds include quick maturing birds, birds with the best stamina, alertness of the birds, birds with good conformation, and capacity to produce many eggs of uniform size.

A good and standard chicken egg should be about two ounce and the weight should range from 55 to 60 grams. Some of the best eggs laying chickens (in terms of the size and number of eggs produced) include White Leghorns and their hybrid strains such as Unichix, Babcock, Poone Pearls, Hisex, and Ranishavers among others. Most of the hybrid strains are known for laying brown or white eggs. However, the nutritional value of the white and brown eggs is usually the same.

Selection for Meat Production

A good meat producing bird should attain a weight of two kilograms at about 10 to 12 weeks of age. Most meat type chickens (broiler) usually attain the required weight much earlier, 7 weeks. Most people prefer birds that are compact because they are usually cheaper and provide adequate meat for a single meal thus preventing wastages. Some of the best meat producing breeds includes the New Hampshire, Cornish, and Plymouth Rocks among other breeds.

Selection for Dual Purpose

Most dual purpose breeds are mainly reared through free range. Therefore, there are certain traits that these birds should possess. They should be hardy enough to withstand the harsh environmental conditions surrounding them, agile in escaping various predators, resistant to various poultry infections, good brooders, and free range freely without much attention and care. Most indigenous poultry breeds are categorized in this section. They are not good meat and egg producer, but with proper management and suitable environmental conditions can improve in production. Poultry breeders can used various exotic breed in grading up these indigenous breeds. Some of the commonly used dual purpose chickens include Australorps and Rhode Island Reds.

CHOICE OF BREEDING BIRDS

When selecting poultry breeds for rearing, it is important to consider their genetic potentials in production. In other words, vigorous and healthy birds are the best for breeders to always select. The following table illustrates various characteristics to check for when selecting a breed for egg laying.

Table 4.1: Summary of various characteristics to check for when to select a breed for egg laying

ITEMS	CHARACTERS
Vigour	• Birds should be active and interested in activities going on around them • They should walk, fly, run, crow, scratch, and present mating interest • Should have a long, broad, and deep body • Good feeds conversion • Should have a bright red and large comb
Head	• Beefy, masculine, phlegmatic and coarse head (traits of a poor egg layer) • Alert and clear cut rugged head (quality of a good egg layer)
Eyes	• Good breed birds should have well placed, bright, and prominent eyes • Good breed should not show signs of depression and physical weakness
Beak	• The beak of a good breeding bird should be relatively short, well curved, and strong • Should not have deformities on the mandibles

Ear lobes	• Good breed should have soft, waxy and properly developed ear lobes • Coarse, wrinkled, and shrunken ear lobes (signs of a poor egg laying breed)
Neck	• A good egg laying breed should have short and stocky neck
Vent and abdomen	• A good egg laying chicken should have pliable pubic bones with about three to four fingers space between them • Good egg laying chickens should have a broad, moist, pliable, and oval shape vent
Body conformation	• Broad and flat back that carries well • Body should be long, broad, deep, and increases in size towards the abdomen • The skin quality should be good (velvety, pliable, and loose) • Should have short and well set apart legs, short nails, and clean and smooth shanks • The keel bone should be relatively long
Temperament	• A good egg laying breed should be active and always alert • Should have good appetite at all time
Moulting	• Moulting season for a good egg laying bird should come late • After moulting, the bird should come back to normalcy in about 4 weeks and should start laying eggs quickly

Source: *https://www.slideshare.net/PakRose1/selection-for-eggs*

When selecting pullets for egg laying, it is important to consider always those that are about 20 weeks old or a little earlier. Studies have shown that pullets that start laying at an early age are usually the best egg producers. For commercially kept egg layers, it is important for the breeders to ensure that 600/0 of all laying birds in specific units is attained daily. In other words, this means that a breeder should collect about six eggs from a unit of 10 pullets every day. Therefore, breeding birds should be selected carefully and all those that are not profitable sold or culled without delay.

CULLING OF POULTRY

Culling refers to elimination of all those birds in poultry flocks that lack the necessary characteristics that they were bred for. These include both meat production traits and egg production traits. Selection and culling are both terms most often used synonymously although they have little connection in poultry breeding. Briefly, selection involves identifying and choosing the best birds in terms of qualities for breeding purposes. On the other hand, culling involves clearing out all unproductive and uneconomical birds in poultry flock. Although a poultry farmer may start breeding with quality stock, customarily there are certain birds that usually fail to grow or produce as it was expected. These birds not only do they affect poultry production negatively but also require similar attention and maintenance as those with good production and normal growth. Culling therefore should be done on these specific birds and should be practiced continuously throughout the year.

Culling can sometimes be a challenging task to the poultry farmers, especially if there are no appropriate ways of identifying and catching the birds. Farmers are advised to use appropriate techniques such as catching crates. These crates are usually partitioned into three sections. All the poultry flocks are supposed to be restrained in their respective houses. Sacks should then be hanged on all the openings of the house in order to darken the house. Catching crates should then be placed outside the pop-hole (poultry exit holes). Birds are driven in to the catching crate, and when it is full, it should be replaced with another until all the required birds are caught.

During the culling process, it is important to handle the birds carefully. Growing flocks and laying birds should not be frightened. When examining the birds, it is important to hold the birds in their natural positions. In other words, one hand should hold the bird while the other is used to handle the bird freely on the wings, legs, and head.

Culling of growing poultry flocks should be done continuously. This helps in preventing occurrence of poultry infections, saving poultry management cost on feeds, and maintaining uniformity in poultry flocks. Birds that often show sickness signs and lack essential vigour should be culled as soon as they are identified in the flock. There are certain signs that every farmer should look for in order to identify a sick bird or one that lack the right vigour. These include dullness of young birds, tendency to huddle around the feeders and drinkers, droopy feathers, lack of normal plumage lustre, and pearly or greyish eyes among other symptoms.

Culling young and growing birds-There may be some birds with feathering problems which extend up to 10 weeks of age. This problem may be as a result of deficiency of certain mineral in poultry feeds or overcrowdings in the house. Sometime, this can be an inherited poultry trait. If such birds are identified in the flocks, they should not be used for breeding purposes. However, they should be culled or marked and later used as table meat when they attain an age of 3 months. Those that possess poorly shaped body should also be culled. Other characteristics that should be included in the culling list are those birds with narrow bodies, long legged birds, those with crooked breast bone, and defective tail carriage among other defects.

Culling of laying flocks-the profits received from eggs laying birds is directly proportional to the amount of eggs produced. Because the cost of purchasing poultry feeds is almost 60% to the egg production total cost, it is highly recommended to identify and immediately cull all those layers with poor laying traits. Those that were good layers but due to certain conditions have stopped laying eggs should also be culled immediately.

EFFECTS OF POULTRY CULLING ON EGG PRODUCTION

Most poultry farmers usually have a perception that total egg production in the flocks will be reduced amicably once some of the laying birds are culled out of the flocks. But the main aim of culling should not be focussed on the total egg production, rather on the profits obtain from the flock based on the total production cost. In other words, farmers should calculate their egg production profits based on the average number of eggs produced by the birds at specific period of time.

Table 4.2: *A suitable form that every poultry breeder can use to record the total number of eggs laid by the birds every day and the total number of dead or culled birds*

SI.No	Date	October		November		December	
		No. of females (first of month)		No. of females (first of month)		No. of females (first of month)	
		No. of eggs	Birds Died culled	No. of eggs	Birds Died culled	No. of eggs	Birds Died culled
1							
2							
3							
4							
5							
Total eggs per month Average eggs/bird/month Birds culled Birds dies							

Source: *https://www.slideshare.net/growelagrovet/poultry-farming-training-course*

There are three ways to calculate the total number of eggs produced by each bird in the flock. These include:

- Average number of eggs produced by each bird every month (monthly average)
- Average number of eggs produced by each bird every day (bird-day average)
- Average number of eggs produced by each bird on a bird housed basis (bird-housed average)

Monthly Average

There are two calculation used in calculating the average number of eggs produced by each bird every month. One ways involves adding the total number of laying birds at the beginning of a month to the total number of laying birds at the end of the month; this total should then be divided by two to get the average number of birds for the month. The total number of laid eggs that month should then be divided by this average number of birds in that month.

For instance, if on the 1st of May there were 100 egg-laying birds, this number of egg-laying birds will be reduced to 90 at the end of the month. To obtain the average number of birds for the month of May, the following formula is used:

$$\text{Average number of birds} = \frac{100+90}{2} = 95$$

If the total number of eggs laid by the birds were 1,800 eggs, the average production of eggs for each bird for the month of May is equal to:

$$= \frac{1,800}{95} = 18.9\, eggs$$

One of the faults in using this calculation method is the lack of information regarding the total number of culled or dead birds during the month.

Bird-Day Average

To get the bird-day average, one should divide the total number of laid eggs within a given period of time by the average number of total birds on hand within that same period of time. This average is usually obtained on a monthly basis.

For instance,

1. $$The\,total\,number\,of\,birds\,by\,May\,1st\,were \quad = 100$$
   $$The\,total\,number\,of\,bird-day\,were\,100x30 \quad = 3,000\,bird\,days$$

2. $$If\,there\,was\,one\,culled\,bird\,on\,May\,5th$$
   $$there\,would\,be\,3,000-25 \quad = 2,975\,bird\,days$$

3. $$If\,there\,was\,one\,died\,bird\,by\,May\,10$$
   $$then\,there\,would\,be\,2,975-20 \quad = 2,955\,bird\,days$$

4. $If\ there\ were\ two\ culled\ birds\ on\ May\ 20$

$then\ there\ would\ be\ 2,955-(10x2) \quad = 2,935\ bird\ days$

5. $If\ there\ were\ two\ dead\ birds\ and\ three\ culled\ birds$

$on\ May\ 25, then\ there\ would\ be\ 2,935-(5x5) \ = 2,910\ bird\ days$

6. $$The\ average\ number\ of\ the\ birds\ in\ May = \frac{2,910}{30} = 97$$

7. $If\ the\ total\ number\ of\ laid\ eggs\ by\ the\ birds\ during\ the\ month\ was\ 1,800,$

$$then\ the\ average\ number\ of\ birds\ per\ bird = \frac{1,800}{97} = 18.55$$

8. $$The\ percentage\ of\ eggs\ production\ per\ the\ month = \frac{1,800}{2,910} x100 = 62\%$$

Bird-Housed Average

To get the bird-housed average in poultry, one should divide the total number of laid eggs within a certain period time by the total number of birds at the start of the period. Based on the example given above, the total number of eggs laid by birds in the month of May was 1,800 and total number of birds in the flock within that month was 100. Therefore:

$$\text{bird} - \text{housed for the month of May} = \frac{1,800}{100} = 18\ per\ bird$$

The bird-housed average is calculated on a yearly basis. The total number of produced eggs within the whole year should be divided by the total number of birds at the beginning of the year. Based on these methods of calculation, a poultry farmer is able to obtain potential income of a flock.

If the need to cull birds in the flock from time to time is reduced with the aim of maintaining production of eggs at an efficient level while reducing the level of mortality in the flock, the average egg production per bird will be enhanced thus resulting to more profits.

INCUBATION AND HATCHING

For the cost-effective poultry production, it is highly advisable to replace the old flock always with a new one after every 2 years (except for the cases of all culled birds, sold birds or dead bird where replacement is always done after every 3 years). This therefore means that one third to one half of the entire birds in poultry flocks should be replaced annually with newly hatched chicks. One of the safest ways of producing new chicks in a poultry farm is through incubating eggs.

Incubation can be described as the practice of reproducing or multiplying poultry stocks through hatching out new chicks from eggs that have already been fertilized, either artificially or naturally. One way of incubating fertilized poultry eggs in setting a broody fowl on the eggs for a certain period of time. This method is referred to as natural incubation. The other commonly used way of hatching fertilized eggs is through artificial method where eggs are set in an incubating machine that provides the same conditions (temperature and humidity). This method is called artificial incubation. Temperature and humidity in modern incubators are regulated using hot air or hot water with electricity or kerosene.

As a poultry breeder, it is important to remember that the incubation period for naturally or artificially incubated eggs take the same time. In other words, this means that artificially incubated eggs of a chicken take the same time to hatch if the eggs way incubated naturally. The following are incubation periods of different birds in poultry flocks.

- Chickens eggs: -21 days
- Geese eggs: 30 days
- Turkeys eggs: 28 days
- Ducks eggs: 28 days
- Guinea fowls eggs: 26 days
- Quails eggs: 18 days

- Pigeons eggs: 18 days
- Pheasant eggs: 24 days

There are several advantages and disadvantages of using either natural ways or artificial methods to incubated eggs in poultry breeding. These advantages and disadvantages are discussed in the table below.

Table 4.3: The differences between the natural incubation and artificial incubation in poultry breeding

SI No	Natural incubation	Artificial incubation
1	This is done using a broody bird	This is usually practice using incubating machines
2	It is relatively cheaper (already fertilized eggs about two to seven days old are set under a broody bird)	It is relatively expensive (already fertilized eggs about two to seven days old are set in an incubation machine)
3	Only a few fertile eggs can be hatched in to new chicks (ten to fifteen eggs) using one bird	A huge number of fertile eggs can be hatched in to new chicks (few hundreds) using an incubating machine
4	Broody poultry birds can be secured easily in the village at a low cost	It is extremely costly to purchase a small scaled incubating machine. Big ones cost thousands of dollars
5	There is no technical expertise and knowledge required because all incubating processes are done by a broody bird	There is the need for trained and qualified personnel. Additionally, an uninterrupted source of heat and humidity must be provided
6	The broody birds helps in regular turning of the eggs	Eggs in the incubating machine are turned mechanically using devices or persons
7	Hatchability is usually high (about 80% to 100 %)	Hatchability is comparatively low (usually 60% to 80%)
8	After the eggs are hatched, the mother bird usually take good care of the newly hatched chicks	Brooder or foster mother is used for rearing the newly hatched chicks.

Source: *https://www.slideshare.net/mohsinjjee/incubation-process-of-chicken*

Natural incubating processes are widely practiced by villagers or traditional poultry keepers, because the farmers are able to collect a small number of fertilized eggs and be able to set the eggs under a broody bird for hatching. It is actually uneconomical to use or set a small number of fertilized eggs in an incubator. Artificial incubation is therefore used by commercial poultry breeder where huge volumes of fertilized eggs are collected over short period of time. Alternatively, several small scaled poultry breeders can collect their eggs individually then combine the eggs and set them in a communal owned incubator for hatching. Proper management of incubator is required. This includes regular supply of necessary temperature and humidity. The required temperature in an incubating machine is usually 37.6 °C (99.7S °F).

Broodiness is usually considered as a maternal character in almost all type of poultry fowls. This trait is usually noticeable and persistent in the indigenous chickens than in any other types of poultry fowls. A broody bird usually has a natural instinct of sitting on eggs and taking good care of newly hatched chicks. Studies have shown that an indigenous chicken usually lay about ten to twelve eggs in small batch, about five to six periods or times in a year. After every period of egg laying, the birds usually become broody for a period of 1 month or more.

There are several factors or indicators a poultry farmer can use to tell a broody bird. A broody chicken or bird usually does not lay eggs. It is also important to note that most of them usually have a tendency of sitting on laying box, they like dark place and warm corners. Other traits of a broody bird include inactivity and dullness. Additionally, most broody birds usually ruffle their feather and tend to sit when they are approached. They also have a character of making clucking noise when disturbed. Most of them usually do not have appetite on feeds and this result to considerably loss of body weight. The sizes of their wattles and combs are usually reduced. The colour of their combs, beaks, and shanks usually turn pale. Loss of lustre on their plumage is also noticeable. High body temperature is also another commonly noticeable characteristic of broody birds mainly because they do not drink water.

When preparing broody bird for sitting on eggs, it is highly advisable to consider those signs mentioned above. Broody hens usually stop laying eggs at least 1 week before sitting on eggs. It is important to ensure that this bird is practically healthy and in perfect physical condition before sitting on eggs for a period not less than 3 weeks. A perfect broody bird should have certain traits such as broad chest, relatively big, short legs, and broad wings. Studies have shown those birds (hens) with long legs and heavy bodies are not the best for sitting on eggs. The bird must be hygienically clean, free from worms, healthy and free from other parasitic infections such as fleas, ticks, lice, and mites. Deworming a week before sitting on eggs is highly advised. It is also necessary to trim the nails on their toes in order to prevent breakage of the eggs during incubation period. A good broody hen should be motherly, this means that they should be able to teach newly hatched chicks on how to scavenge for feeds on the ground and escape in suspicion of a predator. It is also important for the birds to take good care of the newly chicks until they are 8 weeks of age after which they usually separate and go back to laying eggs (this occurs about 12 to 16 weeks after hatching).

Poultry breeders should only use already fertilized eggs for incubation. Fertilization of eggs can be natural (using a male bird) or can be done artificially. It only takes about 3 weeks for chicken fertilized eggs to hatch into new chicks. The germ in a fertile egg usually develops rapidly within the egg to form an embryo. If the aim is to have a successful poultry breeding enterprise, it is highly important to ensure that all details concerning breeding eggs are not overlooked.

Table 4.4: The various traits considered when selecting breeding eggs in poultry

SELECTING THE BEST POULTRY BREEDING EGGS When selecting eggs for breeding, it is important to remember that all laid eggs aren't good for breeding purposes. Therefore, the best and most suitable eggs for hatching should be well selected in order to increase their hatchability, either naturally or artificially. The following criterions are used when selecting these eggs.
• **Fertility** The best and fertile eggs to use for breeding purposes should be fresh and produced by matured and vigorous birds, usually five to seven days after mating. Sometimes eggs collected immediately after mating may not be fertile.

• **Shape and size of eggs** Apart from fertility, eggs used for poultry breeding are supposed to be large enough. This is important because it also determine the size of newly hatched chicks. In other words, breeding eggs should not be too small or too big, depending on the type of breed. The best eggs should be about sixty grams in weight, oval and uniformly shaped.
• **Egg shell** The texture, thickness, and colour of the egg shell should be uniform. This helps in spreading warmth to the eggs uniformly during incubation period. Tinted white eggs should be avoided for breeding purposes. Eggs with dark-brown shells are usually better that light-brown shelled eggs. Poor texture due to lack of vitamin D and calcium may significantly reduce hatchability of the eggs. All eggs with cracked shell should strictly be avoided for breeding purposes.
• **Egg quality** Stale and non-fresh eggs should not be used for hatching purposes. The best eggs for hatching should be collected within a period of 1 week before setting. Two to four days old eggs are the best.
• **Deformity** Poultry breeders should avoid all the eggs with any kinds of deformity. These include cracks, projections, ridges, stains, depressions, or encrustations. Thin-shelled eggs should also be avoided. This is because they may break when turning and may not retain the right amount of warmth or temperature necessary for the formation and development of embryo. On the other hand, think-shelled eggs are also not good because it may be difficult for the chicks to chip it out using their delicate and tender beaks when hatching.
• **Soiled eggs** Poultry farmers are advised not to wash soiled eggs before setting them for hatching. This is because washing usually opens up various pores on the shell and this may interfere with the hatching process of the eggs. If the eggs are slightly soiled, breeders should use a knife to scrap off the dirt. Excessively soiled eggs should be evaded for hatching purposes.

- **Storage**

To get the best hatching results, all collected eggs should be stored in a cool, dry and clean place at all times. Breeding eggs should not be storage for more than three day in extreme hot weather conditions, but in cold and hilly places eggs can be actually be stored for more than ten days. Eggs have life and can actually absorb different odours. It is therefore necessary to avoid storing them near smelly substances like the kerosene, onion, and garlic. When transporting hatching eggs, they should not be shaken too much.

Source: *http://makeitgrowit.com/egg-layers-selection-breeding/*

There are several factors that determine the right time or period to breed poultry. Environmental condition (climate) is one of these factors. Extreme hot or cold climatic condition is not favourable for hatching. This is because it may interfere with the right temperature and humidity necessary for hatching process.

Eggs should be tested well before setting them for hatching, especially if they are collected from different sources. Testing for fertility should be done on the sixth to seventh day after they are laid. All damaged and infertile eggs can be discarded at this stage. One of the best and mostly used techniques for testing breeding eggs is known as eggs candling.

In conclusion, poultry breeding at the village level means mating female birds with male birds. Lack of knowledge may hinder most village poultry farmers from practising fundamental principles of technical poultry breeding. However, with proper guidance, breeders can actually adopt modern poultry breeding techniques. They are able to select the best breeds for their poultry flocks and be able to run a successful and profitable poultry enterprise. When breeding poultry, it is important to always remember that a successful poultry breeding business highly depend on the foundation set for the birds. Culling and other techniques to remove all unprofitable birds from the flocks should be adopted and practiced continuously. This helps in minimizing production cost while maximizing the profits.

CHAPTER 5

EGG LAYERS POULTRY BREEDING

INTRODUCTION

Layer poultry breeding refers to as keeping poultry birds for the purposes of producing eggs commercially. Because of the value attached to these types of breeds, proper management is usually a crucial factor for these types of poultry breeding from time the birds are one day old. Most commercially raised egg laying birds start producing eggs when the reach an age of about 16 to 18 weeks old. Studies have shown that egg production is usually a continuous activity once these birds start producing eggs until they are about 72 to 80 weeks old. Feed conversion rate based on eggs production is usually high. This means that most of them are usually able to produce about a kilogram of eggs by consuming about two kilogram of feeding during their egg production period. Most modern breeders are using different breeds to produce hybrid eggs laying birds.

EGG LAYER BREEDS

Currently, there are different types of poultry breeds that have been produced or development purposely for the production of eggs. One of the best commercial egg laying breeds is the White Leghorn. These birds usually lay white shelled eggs. Moreover, they are usually too flighty. Although there is no nutritional difference between white and brown shelled eggs, many people prefer buying brow-shelled eggs. Because of this, many commercial poultry breeders have developed layers for brown-shelled eggs, with some specifically for the purposes of pasture poultry production. There are also crossed sex-linked eggs laying birds that have been produced. Sex-linking has enabled breeders to develop different types of breeds based on the physical appearance such as plumage colour and body conformation. This new technique of breeding has enhanced the quality of the birds bred and reduced sexing faults.

Some people usually prefer having poultry flocks that are of different breeds. Such flocks are able to produce eggs that have different shell colours. Based on nature, commercially bred poultry birds produce two types of eggs, brown coloured and white coloured eggs. Most white egg laying chickens are relatively small in size. They have a tendency of eating less feeds and produce white-shelled eggs. Some of the examples in this category include Lehman, Isa White, Hi sex, Nikchik, Hi line White, Havard White, Bovanch White, and Sever White among others. For the brown-shelled egg laying chickens are comparatively large in size. They also consume more feeds compared to the white-shelled egg laying chickens. They also produce relatively larger eggs. Among the best known types in this category include the Hi Sex Brown, Lehman Brown, Isa Brown, Hi Line Brown, Gold Line, Bablona Harko, Bablona Tetro, Sever 579, and the Havard Brown among other breeds.

PRODUCTION EXPECTATIONS AND FACTORS AFFECTING EGG PRODUCTION

A female bird is only able to lay one egg per day. There is also a certain period when individual birds stop laying eggs. There are several reasons why birds stop producing egg as before. One of the main reasons is the birds' reproductive system. Studies indicate that the body of a bird has the tendency of forming new eggs shortly after laying an egg. The whole cycle usually takes about twenty six hour for the full formation of an egg. Because the reproduction system of the bird is sensitive to exposure of light, egg laying birds have a tendency of laying eggs late in the day and this actually delay formation of new eggs. This may cause the birds to skip a day or two before laying another egg. There is also evident that most birds actually do not start laying eggs exactly at the same time. It is also important to remember that birds do not continue producing eggs for the same period of time. The figure below clearly indicates a typical curve for the production of eggs in poultry birds. Based on this figure, it is clearly indicated that most birds usually come into production quickly, and after reaching production peaks, the level of egg production is then reduced gradually.

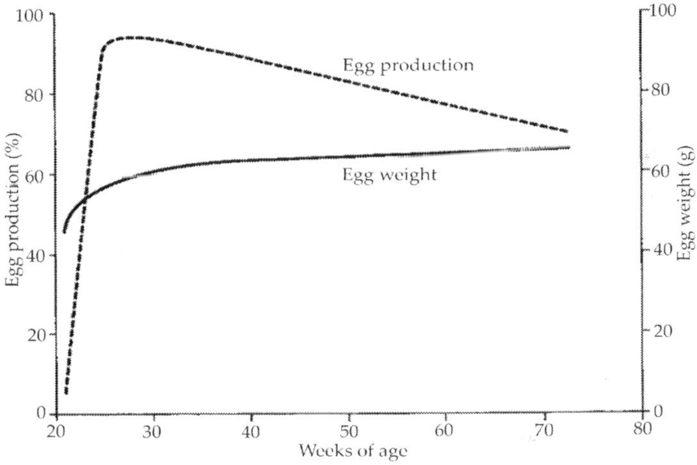

Figure 5.1: Typical curve for the production of eggs in poultry.

Source: *http://articles.extension.org/pages/71004/raising-chickens-for-egg-production.*

The period of time that each bird will lay eggs varies greatly, with most birds taking about 3 to 4 years. Studies indicate that the level of egg production each year is usually less than the previous year. Apart from minimal production, it has also been shown that the quality of the egg shell and size of the eggs are also reduced each year. There are different factors that affect production of eggs in poultry. These factors are discussed below.

Poultry Breed

Some poultry birds have been developed purposely for commercial eggs production. One of the most commonly used breed for the production of eggs is referred to as the White Leghorn. This breed has also been used in grading up some indigenous poultry birds to enhance their egg production. There are other good egg laying breeds that a farmer can also choose from. However, it is highly recommended to always choose commercially bred poultry birds to have maximum production. Studies show that these birds are the best in feed conversion thus translating to more production and high profits.

Management of Pullets

Pullet management is a key factor that influence egg production in poultry flocks. This includes proper light and nutrition management. These two elements are important especially when it comes to the quality and the level of production once the birds start laying the eggs. Proper management of pullet should ensure pullets do not come in production very early. This is important because it may develop some problems during prolapse period, which may lead to various poultry health issues. The birds may also produce small eggs throughout their egg laying cycle. Therefore proper management should start from the time the pullets are one day old. However, for those poultry breeders that purchase pullets that are almost ready to start laying the eggs, they should always remember to ask for their records regarding light and nutrition management. This will help in adjusting subsequent flock management accordingly.

Management of Light

Most chickens are commonly referred to as long-seasoned birds. This means that their egg production is directly influence by the length of daylight. In other words, most birds will begin producing eggs when day light hours are more. Naturally, most day-old chicks are reared on a 24-hour schedule for the first few days in order to familiarize them with feeds and water. The number of lighting hours should then be reduced once the chicks are aware of their feeding and water drinking points. Indoor reared chicks should be limited to about eight to ten hours of light per day. When the chicks reach their pullet stage, the number of light hours per day should be increased to about fourteen hour of light per day. This exposure to light helps in stimulating the pullets to actually start laying the eggs. To help the pullets remain in productive, it is advisable to slowly increase the length of light hours per day to about sixteen hours. To most poultry breeders, this would actually mean supplementing the natural day light with some artificial lighting.

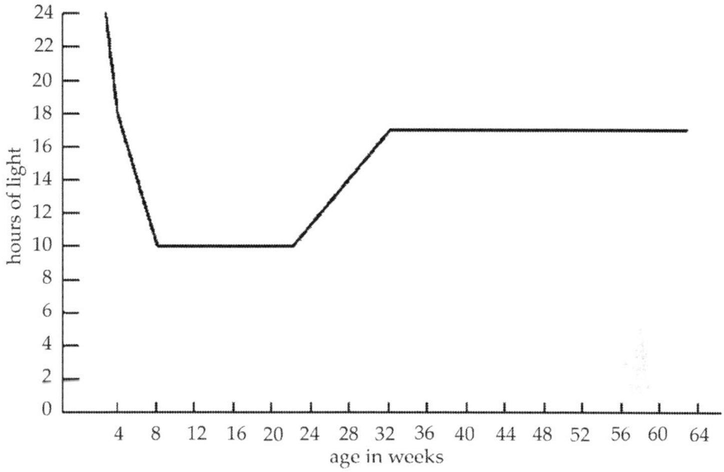

Figure 5.2: Effects of hours of light on egg production in poultry.

Source: *http://www.fao.org/docrep/005/Y4628E/y4628e03.htm*

One strategy of adding the number of light hours to the eggs laying birds is to have a start/stop timer, which will help in switching the light on late in the evening before sun set and early in the morning before sunrise. Alternatively, a farmer can have a light sensing switch that ensures that lighting bulbs do not switch on when there is still natural light in the house of the pullets.

Nutrition

Well-balanced poultry diet is recommended for birds of all ages. Most commercially prepared poultry feeds are good because they contain all nutrients combinations for better growth and development of birds. Some poultry breeders also have a tendency of mixing their own poultry feeds using cheaper grains, and this has always contributed to poor poultry feeding thus resulting to various poultry health and production issues due to deficiencies of essential nutrients in their body. Deficiencies of crucial nutrients can affect the growth and development of most pullets and their production levels.

Farmers and poultry breeders are advised not to give their growing or egg laying pullets meat type poultry feeds. This is because these feeds may not supply the required nutrients to the bodies of pullets. Similarly, growing chicks should not be fed with layers feeds. Layers feeds are usually rich in calcium, which is required by the birds for eggshell formation. High level of calcium is however very harmful to most non-laying chickens.

Space Allowances

Egg laying birds should be provided with adequate space to enhance their egg productivity. The amount of space on the floor required by each bird is directly influenced by the type of housing and the size of the bird, which is also related to the type and breed of the chicken. The recommended floor space should be two square feet per bird, whereas the minimum space should be one and half square feet per bird. Including perches in the house of the birds may help in facilitating housing for large number of birds. Apart from housing element, perches helps in accumulation of poultry droppings in one place thus easier cleaning. The welfare of birds that enjoy perching is also provided. Sometimes egg laying poultry birds are also allowed to have access to the outside space. Therefore, the amount of space will depend on whether there is pasture for the birds to scavenge for food or the space is only meant for outdoor access. If poultry are allowed to have outdoor access, it is advisable to be aware of various poultry predators, both in the air and on ground.

Climatic Conditions

Production of eggs is usually influenced by the climatic conditions of a place. The optimal amount of temperature for egg laying usually ranges from 11 to 26 °C. Apart from temperature, humidity is also a crucial element that also influences egg production in poultry. Usually, humidity level of about 75% and above will lead to a reduction in production of eggs.

When the environmental temperature increases over 28 °C, the egg production and quality are usually affected negatively. Based on studies, it has been revealed that increase in seasonal temperature can actually lead to about 10% decrease in eggs production.

Table 5.1: Illustrating how eggs production is influenced by temperature effects

Temperature (°C)	**Effects of temperatures**
11–26	Production of egg is usually good
26–28	Most birds usually have a tendency of losing appetite to feeds
28–32	Both water intake and consumption of feeds are usually reduced amicably. This results to reduction in egg production, reduced size of the eggs, and thin-shelled eggs
32–35	Most birds are seen panting and distressed
35–40	Prostration due to excess heat is observed. Breeders must adopt new house cooling measures in order to save their flocks
≥ 40	High mortality resulting from excess heating in the house

Source: *http://www.fao.org/docrep/005/Y4628E/y4628e03.htm*

Poultry disease management-effective and efficiency poultry management practises are significant in enhancing production and consequently increase profits. This entails not only proper poultry housing and efficient feeding but also proper poultry rearing and disease control practises. Diseases in poultry production cause high eggs production losses. Some of main poultry diseases include bacterial, protozoan, viral, nutritional, and fungal disease. Additionally, there are several parasites (both internal and external) that also result to reductions in poultry production. Administering vaccines is highly required. This can be done through injection, eye drops and drinking water intake. Hygienic and clean poultry houses and environments may help in eliminating up to about 80% of the poultry disease occurrences.

Mortality rate- mortality rate in poultry can increase due to several factors such as high temperature, predation, and poultry disease infections. Mortality of young chickens (about 8 weeks of age) usually ranges from four to six percent, while that of growing chickens (about 8 to 20 weeks of age) usually range from 10 to 15%. Additionally, mortality of egg laying birds (about 20 to 70 weeks of age) is usually about 12%. In average, the annual mortality of chickens usually range from 20 to 25.

Other factors that influence egg production in poultry include culling (removal of all unproductive and/or sick birds form the flock), provision of adequate egg laying boxes, and the age of the birds.

EGGS PRODUCTION CYCLE FOR LAYERS

Most egg laying poultry birds usually start laying eggs when they reach an age of 20 to 24 weeks. They continue laying eggs for about 52 weeks on average. Egg production decreases when the birds are approaching moulting period. Typically, the egg production cycles for most commercial egg layers lasts for about 17 months (72 weeks) and usually happen in three distinctive phases.

The first phase involves brooding period for the young chicks, usually from first day until the age of 8 weeks. During this period, young chicks are kept or reared in separate house from eggs laying poultry birds (brooder houses).

The second phase usually involves grower period. This lasts for a period of 3 months. Grower chickens can either continue living in the brooder house or can be reared separately from newly and young chicks. It is highly advisable to provide proper care to the grower chickens particularly when they are about 17 to 20 weeks of age. This is because most of their reproductive organs grow and develop during this phase.

The third phase involves the layers birds. Grower chickens are moved from their houses to the layers houses when they attain an age of 18 weeks. This is done to prepare them for eggs laying cycle. Typically, most poultry layers usually start laying eggs at the age of 21 weeks, and this continues until they attain an age of about 72 weeks.

Production Planning

Averagely, a bird lay one egg in a day. Additionally, poultry birds start laying eggs at different ages. For constant egg production to meet the demand in the market, proper planning for poultry production is important. Poultry breeders can adopt the following schedule that indicates proper scheduling for poultry eggs production based on average satisfactory levels for productive birds.

Table 5.2: Illustrating a production schedule intemperate climate (100 birds)

Age of flock (in weeks)	% of flock laying	No. of birds laying	No. of eggs produced per week
21	5	5	20
22	10	10	40
23	18	18	72
24	34	34	136
25	52	52	208
26	65	65	260
27	74	74	296
28	84	84	336
29	88	88	352
30	92	92	368
31	94	94	376
32 - 39	88	88	352
40 - 47	83	83	332
48 - 59	77	77	308
60 - 64	73	73	292
65 - 70	70	70	280

Source: *http://www.fao.org/docrep/005/Y4628E/y4628e03.htm*

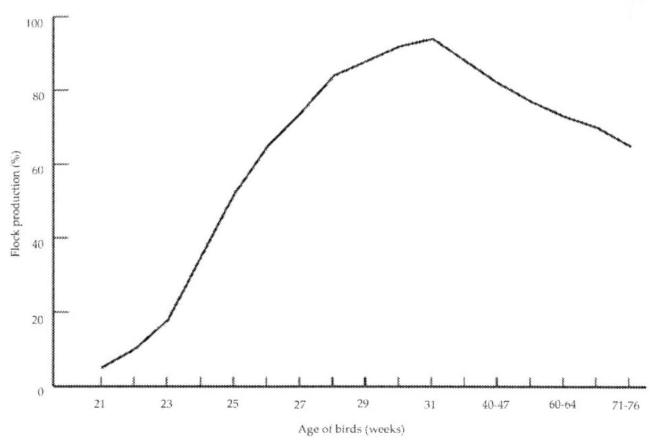

Figure 5.3: Percentage of productive laying flock over period of time.

Source: *http://www.fao.org/docrep/005/Y4628E/y4628e03.htm*

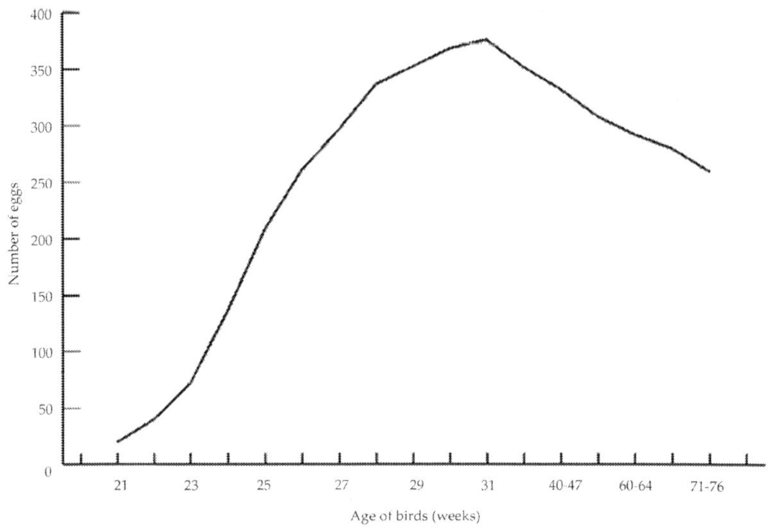

Figure 5.4: Number of eggs produced over a period of time.

Source: *http://www.fao.org/docrep/005/Y4628E/y4628e03.htm*

Most commercially improved chickens (hybrids) produce about 180 to 200 eggs annually in areas that have hot and humid climatic conditions. It has also been revealed that birds in areas with moderate climatic conditions can produce from about 250 to 300 eggs annually. This is well illustrated in the table above.

In the first column of the table indicates the age of the birds in the flock while the second column indicates the percentage of birds that lay eggs within that week of their ages. The third column indicates the actual number of birds laying eggs within that week. Lastly, the fourth column shows the actual number of produced eggs. On average, one bird lays about 208 eggs within a period of 12 months. This average represent weekly egg production rate of about four eggs in each bird. At the age of 21 weeks, it is shown that only twenty birds out of a hundred produced eggs. Production of egg is seen increasing rapidly until 30 weeks of age but start falling after. If egg production at the age of 72 weeks decreases to less than sixty five percent, it become uneconomical to keep on retaining the birds. At this moment, it is highly advisable for the poultry keepers to consider culling unproductive birds for sales. This helps in reducing production cost while maximizing poultry profits. However, if the price of the eggs is high, it advisable to delay culling the birds to a point where about 45% of the birds are still producing eggs (78 weeks of age).

Based on studies, it has been revealed that production of eggs in poultry is directly influenced by planning for the production cost. This involves both the number of produced eggs in a specific period of time and also continuously hatching eggs in to new chicks in order to replace those birds culled as they approach their unproductive periods. A poultry famer can adopt a simple chart like the one shown below when hatching is been done so that to maintain and have a constant egg production in the farm.

Table 5.3: A sample of a production planning chart

Layer flocks	0	11	21	31	41	51	61	71	81
	(...................... time in weeks........................)								
1st layers	Born		Lay						
2nd layers			Born		Lay				
3rd layers				Born		Lay			

Source: *http://www.fao.org/docrep/005/Y4628E/y4628e03.htm*

As shown in the chart above, the first generation of the layers was hatched at the age of 0 week to start laying eggs at the age of 21 weeks. The second batch or generation should be hatched at around 21 weeks so that they can be ready to start laying eggs at 41 weeks when the first generation approach their unproductive stage. In other words, this planning chart indicates the aid of having laying birds that are of different age groups.

Biosecurity and proper hygiene in poultry houses and surroundings should be maintained in order to control and prevent occurrence of various poultry diseases that may affect their productivity. It is also important to ensure that there are no more than three different age groups of the flocks present at the same time. Average mortality rate of egg laying chickens is usually 20 to 25. Therefore, if a poultry farmer for instance wants 100 birds to lay eggs on a daily basis, he or she must plan to purchase or hatch about 120 to 125 chicks.

PRODUCTION COSTS AND PROFITS

Record keeping is highly advisable especially for the incurred costs of eggs production from day one until the birds are culled. Most costs of production are usually covered using the sales obtained from eggs. To determine the actual profits from an eggs production enterprise, poultry keepers should get the difference between the sales obtained from eggs and the production costs incurred throughout for various production operations.

Brooder-grower stage: Production cost should include all the operations costs incurred during the brooder stage and grower stages of the chicken when egg production is usually zero. The brooder stage usually lasts for about 5 months starting from day one to the 20th week. These costs are well illustrated in the table below.

Table 5.4: A sample of expenses for rearing chickens

Costs	**US$**
Chicks (total number of chicks multiplied by price per chick)	
Feed (total kg of feed multiplied by price per kg)	
Housing	
Equipment	
Labour	
Vaccinations	
Mortality	
Loan	
Various	
Total costs	

	US$
a) Eggs sold	
b) Feed used	
c) Rearing costs	
a minus b and c =	

Source: *http://www.fao.org/docrep/005/Y4628E/y4628e03.htm*

Layers production cost-Productions for egg laying birds should be done on a daily, weekly or monthly basis. However, one of the most useful calculations is that is done at the end of egg production cycle. Poultry farmers can actually determine estimated production costs and relatively profits or loses based on the daily, weekly or monthly calculation done. The key issue that bother most poultry farmers is determining whether or not incomes from egg sales can actually cover all production costs on poultry feeds and other maintenance expenses. Generally, the costs on poultry feeds are usually projected to about seventy five percent of the total costs for egg production. Comparing weekly or monthly production costs on various operations involved can actually give a clear indication of profitability or loss for the poultry business. In other words, in order to get the total value of profit or losses made, farmers should subtract the total production costs from the total sales from the eggs, either on a weekly or monthly basis. Additionally, total production costs (the total of cash spent prior to the time the birds started laying eggs) should be repaid. Calculations for poultry egg laying cycle (52 weeks) should be made accurately in order to enable the breeders evaluate and determine profitability of an egg laying poultry business.

Production Costs

When estimating the production costs for a poultry egg laying cycle, there are several expenditure that must be included in the calculation. These include the rearing production costs from the brooders stage until when the growing chicks become layers. There is also the cost of housing the birds which should include building and maintenance costs for the egg laying and brooders houses. Equipment costs should also be calculated on various miscellaneous items that are purchased during the production processes such as feeders, drinkers, and buckets among other items. Another production cost should involve labour costs incurred during management of birds in the house. Vaccination costs should include all the costs incurred to purchases poultry medication and veterinary expenses. There are also the mortality costs and others miscellaneous costs on water and lighting in poultry house.

Income

The revenue obtained from a poultry egg laying venture should be calculated from various operations such as the total value for egg sales, sales of all birds culled after they become unproductive, and where appropriate, the sales of poultry manure.

When deciding to start a poultry egg production venture, one should always have enough initial capital to start the venture; however, various operations production operations should be funded using the sales obtained from the eggs produced before the end of egg laying cycle. The sales from eggs should be sufficient to run the business when egg laying cycle is at 30 to 31 weeks when the production is at its peak.

Table 5.5: Costs and income for production cycle

Costs	**US$**
Rearing (carried forward from table 4)	
Houses	
Equipment	
Feed	
Labour	
Vaccinations	
Mortality	
Various expenses	
Total costs	
Income	
Sale of eggs	
Sale of culled birds	
(Sale of manure)	
Total income	
Profit	

Source: *http://www.fao.org/docrep/005/Y4628E/y4628e03.htm*

SELECTING BIRDS FOR EGGS PRODUCTION

Poultry farmers and breeder should have basic information on how to identify the best birds that will surpass production of eggs. When selecting the birds, it is important to compare and select egg laying breeds based on their characteristics on same sex, breed, and relative ages (for instance, pullets should not be compared to mature birds). When identifying breeding birds, it is important to look at various physical features of the birds. These include the wattle and comb which should be large and soft. Good egg layer pullets should have stretchable and velvety skins. They should have a relatively large, pliable and soft abdomen. The distance between the pelvic arch should be about three to four fingers apart. The rear tip of the keel bone should also be large in size. The vent of a good egg layers should also be wide and moist.

On the contrary, poor egg layers should have shrivelled, hard and short wattles and combs; the vent should be small, dry and puckered; the distance between the rear tip of keel bone and pelvic arch should be short; bad layers have coarse texture and tight skin; and most of them have hard and firm abdomen.

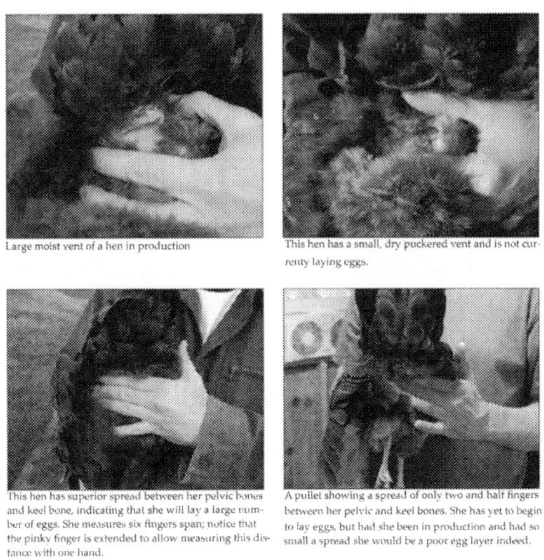

Figure 5.5: Illustrating various features to check for when selecting egg laying birds.

Source: *https://vomitingchicken.com/solve-the-mystery-which-hens-are-laying/*

Moulting ability is another factor that poultry farmers should use when selecting their egg laying poultry. It is important to consider those breeds that moult all their feathers at once than those that take a longer time to shade off. Birds with fast moulting ability are profitable than those that take more time because they are able to resume laying eggs much faster. Studies have shown that most birds that have a late moulting usually become unproductive after about 6 to 8 weeks of egg laying.

This pullet has dropped a few feathers at a time - a sign of slow molting.

This hen is growing all of her feathers in at the same time - a sign of fast molting.

Figure 5.6: Moulting signs to check for when selecting egg laying chickens.

Source: *http://countrysidenetwork.com/daily/poultry/feed-health/when-do-chickens-molt-video/*

When looking for a good egg laying bird, it important to look at the body capacity. The space between the legs should be wide. Productive breed also have full and well-rounded breast. The rear part should be deeper than the front part. This is an advantageous especially in allowing more room for egg production organs. It has also been revealed that those birds that start laying eggs early are often good layers. In other words, pullets that start laying eggs when they are about 6 to 7 months end up been excellent egg layers.

FEEDING EGG LAYERS POULTRY

There are numerous companies all-over the world that formulate and produce poultry feeds for eggs laying birds. A poultry farmer can also decide to make own poultry feeds or purchase some from local markets. When formulating own poultry feeds or purchasing some from the market, it is highly advisable to ensure that the feeds contain all the necessary nutritional elements that egg laying poultry birds require for the production of health eggs. Minerals, vitamins and proteins are essential elements for egg laying poultry birds, whether in the early stages of growth or in egg laying stages. These elements are important because they influence the quality of produced eggs in one way or the other. They are also essential in the fertility of the egg layers and the overall health of the birds.

Two weeks after new chicks are hatched, they should be given poultry feeds that contain about two percent of calcium. Daily monitoring on the feeding behaviour of the chicks is essential. Farmers are advised to feed the young birds with starter feeds for a period of 8 weeks if there are signs of not gaining any weight on the birds. Growing chickens should be served with at least three meals every day until they are about 18 weeks of age. Once the birds start to lay the eggs, demand for more and more feeds usually go up. During laying stage, poultry layers should be fed with well balance feeds. The quantity should be determined based on their body weight and general age of the birds. However, the amount of poultry feeds should not be lowered especially when egg production is at its peak.

Water management for the laying poultry birds is also a very important area to always emphasize. The health and proper egg production is determined by the supply of fresh, clean and pure drinking water. Water supply should be adequate based on the demand for the eggs laying poultry birds. Proper positioning should be done within the poultry house where all birds can access the water at ease. During hot weather seasons, drinking water should be cold; while during winter seasons and cold weather slight warm water should be supplied for the eggs layers.

Based on the species and the age of the chickens, the weight of the birds can actually be control and regulated using the kind of poultry feeds provided. Poultry feeds for layers should contain adequate vitamins, calcium, amino acids, and phosphorus among other essential mineral substances. Drinking water for poultry should be purified using essential substances such as Aquacure.

SEQUENCES FOR EGG LAYING POULTRY BIRDS

There is certain procedure that is followed for egg laying poultry birds. These sequences start by hatching fertile eggs and turning young female chicks into commercial pullets for the production of eggs. This stage is usually referred to as the rearing stage. There are several methods or systems used for rearing the young chicks. Most farmers usually adopt or choose to rear their chicks in a house with well littered floor. This help in warming the house and preventing dampness in the house. Some farmers also choose to rear their pullets in cages with wired-floor.

Brooding stage is a very essential stage for poultry egg layers. This is the stage that determines whether newly hatched chicks will actually survive to the grower and layers stages. This stage usually lasts for about 6 weeks, and it is influenced by environment temperature when the chicks are actually able to regulate and control their body temperature on their own. From the first day until the 6th weeks they should be given starter poultry feeds in order to supply adequate protein (usually nineteen percent) and enough energy for the growth of their bodies.

During growers stage (6 to 20 weeks of age) the birds still need protective care from the farmers. During extreme cold weather seasons, the birds should be provided with artificial heat supply. It is also advisable to give the growing pullets the grower poultry feeds which should contain about 15 to 17% protein and about seven percent less energy compared to the starter poultry feeds.

Trimming of the beaks and several vaccinations are also done during this stage of growing. This helps in preparing the young pullets for their egg laying adult stage. Studies have revealed that anything that may hider or limit the growth of pullets may lead to negative consequences in terms of ability to lay eggs adequately. However, farmers should also regulate the feeding program for the growing pullets in order for them not to gain excessive weight. It is therefore important to restrict feed supply to the poultry birds at this stage in order to modulate their growth rate. This may also help in saving the feeds and increase production of eggs when the pullets reach their production stage. Weekly body weighing is essential for the growing birds in order to restrict weight of the birds through feeding and also determine the growing pattern of the birds.

When the pullets are about 16 to 18 weeks of age, they should be move from their grower houses to layers quarters before they actually become adult egg laying hens. This helps settling them before the start of egg production stage. It is important to handle each bird with care in order to prevent injuring them. During this transition period from pullets to adult egg layers, proper diet should be given to them.

From twenty one up to 72 weeks of age, it is the most important stage of egg production. Proper feeding is required during this state to keep the performance of egg layers high. The temperature in their house or laying coops should range from 21–28 °C. Proper and regular monitoring on their performance and health should be done often. For brown egg strains, the birds should be fed about a hundred and twenty grams of feeds per bird on a daily basis, while tinted egg production birds should be fed about a hundred and five grams of feeds per bird on a daily basis.

The quantity and quality of layers feeds may vary based on the production level of the birds. Most layers usually require more feeds prior and when their egg production is at the peak. It is referred to as "phase feeding". Farmers are therefore advised to adjust the feed according to the needs of the birds.

EGG COLLECTION, PACKAGING, STORAGE AND TRANSPORT

Egg collection mechanism highly depends on the number of birds and those that are actually laying the eggs. Modern commercial egg production business practise extensive egg laying and this need more specialized mechanism for collecting the eggs especially during peak laying seasons. Once the eggs are collected, they should be moved from the pen house to eggs room for grading and checking damaged and cracked eggs. Some eggs may be broken in order to check the quality of eggs from inside. Once grading is over, packaging of the eggs should be done ready for marketing.

There are numerous reasons why egg production businesses should pack their eggs before delivering them to buyers. Packaging can be considered as the art and science of preparing farm produce for storage, transport and finally sale to the end users. Packaging helps in protecting the eggs from tainting, moisture loss, deterioration due to temperatures, microorganisms such as bacteria, natural predators, and from cracking when been transported, stored, or handled.

Since eggs have lives in them, they need to breathe. Because of this, proper packing material should be chosen appropriately to allow exchange of gas in and out of the eggs. It is possible to reuse packaging material, but care should be taken not to taint and contaminate the material. It is also important to protect the packaging material from odours, damage, and dirty. There are several factors that must be considered when packaging eggs for sale. Poultry farmers should seek for appropriate information regarding storage facilities, quality maintenance, transport type used, distance to be covered, environmental/climatic conditions, costs and time involved before the eggs reach the market.

Once the eggs are laid by the birds, they should be stored in a proper way. There are several conditions that every poultry farmer should adhere to. These include storing the eggs in a clean, dry, and cool place. Cleanliness and hygiene of the packaging material should also be maintained. Temperature and humidity in storage room should be regulated in order to prevent loss of moisture in the eggs due to evaporation. All tainting materials and objects in storage rooms should be removed and cleaned using odourless sanitizers and detergents. Proper and fresh air circulation in the room is also encouraged. This enables the eggs to breath freely without any contamination on them.

For successful transportation of eggs from the farm to the market, there are three key requirements that should be met. These include; all material used for egg packaging should be selected well in order to prevent and protect eggs against mechanical damages. The second requirement involves taking care when handling and transporting the eggs to the market. Lastly, when transporting the eggs to the market, it is highly advisable not to expose them to harsh climatic conditions as well as contaminating material that may reduce the quality of eggs.

The required temperature during storage and during transportation is determined by prevailing climatic conditions and the time take to actually transport the eggs from the farm to the market. The following table illustrates the recommended temperature required for the transportation of eggs to the market.

Table 5.6: The recommended temperatures for eggs loading and transporting

	Transport over 2 or 3 days	Transport over 5 or 6 days
Maximum on loading	+6° C	+3°C
Recommended for transport	-1° to + 3° C	-1° to + 1°C
Acceptable for transport	1° to + 6° C	1° to + 3°C

Source: *http://www.fao.org/docrep/005/Y4392E/y4392e0f.htm*

Extra care is needed to prevent shaking the eggs when transporting to the market, especially when the roads are not up to standard. Containers used for egg packaging should be stacked and tightly tied securely in order to prevent movement during transportation. It is important to cover the top to protect and prevent direct sunlight, extreme cold, and rain where applicable.

CONCLUSION

Egg production it an important entity in agricultural industry of any country. It helps in both promoting trade and supply of food in the country. To increase production of eggs in poultry farms, every breeder has a duty to properly select the right breeds for egg laying. Proper management is also necessary in the egg production business. Farmers are also advised to cull their poultry birds especially when they stop producing eggs.

There are several factors that cause bird to stop laying the eggs. Most chickens stop producing eggs when their egg laying cycle is complete. Once this cycle is complete, most birds usually undergo a moulting period (feather losing period). This period usually start after a10-week period of egg laying. Lack of adequate and well balanced poultry diet is another factor that causes birds to stop laying eggs. Birds should be fed well with all essential mineral and water for them to maintain their productivity.

Some farmers feed their poultry birds with improper feeds such as kitchen scrap material and scratch grains instead of appropriate feeds for egg layers. This causes them to lack essential nutrients in the body, thus resulting to moulting and stop producing eggs. When birds, especially layers lack essential minerals in their bodies, it is common to seen them pecking their feathers and producing less or no eggs at all.

Decrease in the number of day light is another factor that causes egg laying birds to stop producing eggs. Farmers are encouraged to supplement the natural light with artificial light especially during the months of winter. Both internal and external parasites also contribute greatly to the reduction in eggs production in poultry birds. Most of them cause a lot of distress to the birds. Heavy internal infestations may actually results to severe damages on the internal organs such as digestive tract thus leading to poor egg production. Studies have also shown that most bacterial, viral, and fungal diseases are major causes for birds to stop laying eggs in poultry farm. These diseases cause production of poor and abnormal eggs with thin and sometimes without shells.

CHAPTER 6

MEAT TYPE POULTRY BREEDING

INTRODUCTION

Artificial poultry selection methods have being applied in development of modern meat type birds. The greatest milestone in application of genetic in poultry breeding was realised in twentieth century. Modern meat type (broiler) chicken breeding is practised using two key steps: the highest level step, which involves controlling and owning pure-breeding lines by various commercial breeders. Based on these lines, birds are subjected to full identification process. Pure-bred birds are then subjected to mass poultry selection based on their traits and characters. Since 1950s, broiler poultry breeders have always being practising their selection on the growth rate of the birds. Modern breeders are also emphasizing their selection criterion on the breast meat of the birds, feed conversion efficiency, and liveability of the birds. Commercial broiler breeders are using the grandparent birds in developing their stocks through crossbreeding.

The other step involves cross-breeding artificially selected birds to obtain the hybrids. These birds are then distributed to various production companies for broiler chickens breeding and productions. Poultry breeders have achieved a lot in the use of artificial selection processes for the last few decades. They have actually shifted from the use of mass selection criteria to the use of hybridisation and pedigree charts among other techniques such as artificial insemination and selection indices in their breeding programs. Based on various technological advancements that have being realised in poultry production industry, especially in production of meat- type chicken breeds, studies have shown that the growth rate of broiler chickens have increased by 400% between 1950 and 2005.

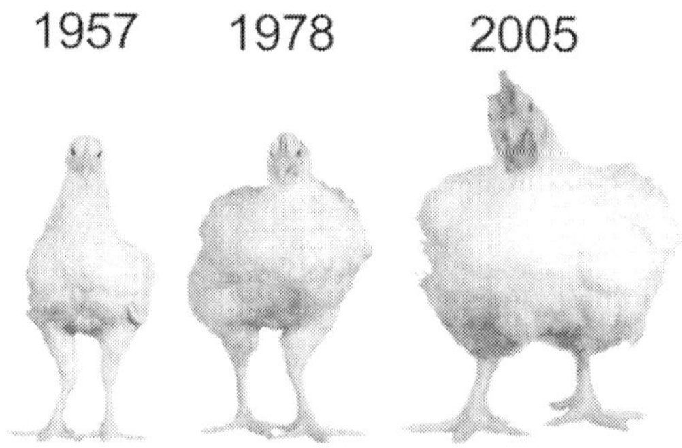

Figure 6.1: Commercially reared broiler genotype produced in 1950s, 1970s, and 2005.

Source: *http://infra-ltd.blogspot.co.ke/2015/03/look-at-what-our-obsession-with-white.html*

These birds are of the same age (56 days of age) and have been fed on an identical modern broiler diet ration. They weigh 9.5 grams, 2,808 grams, and 4,202 grams respectively.

Most broiler poultry birds get their metabolic functions and growth energy from various feeds given to them. Broiler feeds should have all the essential elements such as fat, proteins and carbohydrates that promote growth and development.

BROILER GROWTH AND MANAGEMENT

The most crucial process in production of meat type birds in the farm involves selecting and breeding the best breeds. Broiler chickens should be selected based on their growth rate. This means that their feed conversion to weight gain should be high in order to reach the market weight within the shortest time possible. Male broiler birds have the highest growth rate than female broilers. They weigh about 2 kilograms (live-weight) within a period of 6 weeks of age. The growth rate of female is slightly lower and this is somehow advantageous because different consumers prefer broiler chickens of various weight and size.

Most broiler producing breeders prefer the white feathered breeds because of their clean-looking bodies once they are processed. Because utilisation of feeds is usually the main focus in production of broilers, proper and high quality poultry feeds should be provided. The idea of poultry broiler production should be focused on how the birds consume the feeds and be able to convert those feeds rapidly in to body meat. Therefore, it is the responsibility of each broiler farmer to be conscious about management of feeds.

Broiler feeds should provide the essential minerals and vitamins such as A, B2, D3, B12, and vitamin K. Most of these nutrients are sufficient in various ingredient used in processing poultry feeds, such as cereal grains, micro and macro minerals, animal proteins and fats, vegetable fats and proteins, and various vitamin premixes.

There are different types of broiler feeds. These include pre-starter feeds, which is given from the first to the tenth day of age. The aim of giving the bird pre-starter feeds is to establish and enhance their appetite in order to achieve maximum growth rate within the shortest time possible. Most broiler farmers should target attaining a body weight of about 170 grams and above within a period of seven days for their chickens. Since pre-starter feeds usually represent the smallest proportion based on the total cost of the feeds used, farmers should always make proper decision whether formulation of this pre-starter feeds should be based on the cost or performance and profitability of the birds.

When purchasing the feeds, it is important to always consider the level of digestible amino acid in them. The total fat level of pre-starter broiler feeds should be less than five percent. It is also advisable to avoid feeds that have high levels of saturated fats.

The grower feeds should be given from 11 to 24 days of age. When transiting the type of feed for the broilers birds (from starter feeds to grower feeds), emphases should be given to the textures of the feeds from crumbs to pellets. To help in maximizing performance of the broiler birds, there is the need for continuous improvement on the quality of feeds given to them.

The last type of broiler feeds that should be given to the birds is referred to as the finisher. It should be given from twenty fifth day until the marketing of the broiler birds. Finisher feeds for the broiler usually account for the most expenses used in production of these meat type birds. It is therefore important to apply various economic principles especially when formulating these feeds. Most body changes and composition such as fat deposition and breast meat formation usually happen during this stage of broiler production. Proper consideration on the feeds nutrients is therefore advised.

Table 6.1: The specifications of various nutrients for the newly hatched broiler chicks grown to about 4-6 lb. live-weight

		Starter		Grower		Finisher	
Age fed	days	0-10		11-24		25-slaughter	
Energy	kcal	3025		3150		3200	
	MJ	12.65		13.20		13.40	
AMINO ACIDS		Total	Digest[1]	Total	Digest[1]	Total	Digest[1]
Lysine	%	1.43	1.27	1.24	1.10	1.09	0.97
Methionine & Cystine	%	1.07	0.94	0.95	0.84	0.86	0.76
Methionine	%	0.51	0.47	0.45	0.42	0.41	0.38
Threonine	%	0.94	0.83	0.83	0.73	0.74	0.65
Valine	%	1.09	0.95	0.96	0.84	0.86	0.75
iso-Leucine	%	0.97	0.85	0.85	0.75	0.76	0.67
Arginine	%	1.45	1.31	1.27	1.14	1.13	1.02
Tryptophan	%	0.24	0.20	0.20	0.18	0.18	0.16
Crude Protein	%	22-25		21-23		19-23	
For optimal portions margin it is recommended that amino acid density be increased up to 5% in all diets							
MINERALS							
Calcium	%	1.05		0.90		0.85	
Available Phosphorus	%	0.50		0.45		0.42	
Magnesium	%	0.05-0.50		0.05-0.50		0.05-0.50	
Sodium	%	0.16-0.23		0.16-0.23		0.16-0.20	
Chloride	%	0.16-0.23		0.16-0.23		0.16-0.23	
Potassium	%	0.40-1.00		0.40-0.90		0.40-0.90	

Source: *http://spesfeed.com/2014/04/nutritional-strategies-and-opportunities-for-the-broiler-industry-in-south-africa/*

Broilers usually feed and drink water throughout the day and night. It is therefore important as a breeder to ensure that there is an adequate feed and water supply at all times. During the winter, supplied water should be clean and warm, while during the summer season clean and cold water is recommended for the birds.

Poultry diseases should be prevented and controlled effectively. Disinfection should be done on a regular basis in order to control the occurrence and the spread of various poultry diseases. Other preventive measures should include starting the broiler production business with disease free stocks, preventing poultry feeds contamination with aflatoxins, covering broiler house with litter in order to minimize disease problems, provision of foot-bath at the house entrance, and providing enough fresh and clean water to the birds at all times. The birds should be vaccinated appropriately within the required timelines. This helps in preventing various bacterial and viral poultry diseases.

SELECTING MEAT TYPE BREEDS

Unlike in the case of egg laying birds, broiler choices are usually limited. Some of the commonly used breeds include the Cornish X; other broiler hybrids which come in different colours such as silver, barred, or reds; and dual purposes broiler type such as barred rocks and buffs varieties.

There are several traits that can be used when selecting a good broiler chicken either for breeding purposes or for business. Most of the best meat type birds have a wide skull. This can be used as a perfect indicator for good growth potential. If the skull of the birds is narrow, the body of the birds will definitely be narrow too. Therefore, a medium to large skull width is a good indication for a perfect egg layer, while a large to extra-large skulls indicates a perfect bird for meat production.

In both photos the bird on the right has a wide skull and the one on the left a narrow skull. A wide skull is a good indicator of a good skeleton and good rate of grwoth.

Figure 6.2: Skull features to check for when selecting meat type chickens.

Source: *http://queenofthereddoublewide.com/tag/raising-chickens/*

The heart girth of the bird should be large enough. A good and large heart girth indicates sufficient space for various internal organs of the broiler birds, thus maximizing the potentiality of birds growth and development. For more accurate evaluation on the heart girth of the birds, the legs should be held pointing at the rear side of the body. One should then assess the girth by placing the finger on each side of the birds' ribs around the area where the wings are attached to the body.

Measuring heart girth on live bird.

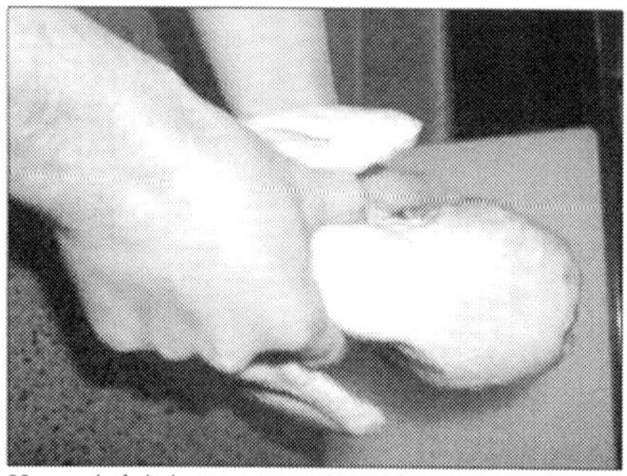

Heart girth is imporyant as it yield space for the heart and lungs.

Figure 6.3: Meat type chicken selection based on heart girth.

Source: *http://queenofthereddoublewide.com/tag/raising-chickens/*

The back of a good broiler bird should be flat, long, and wide. These features actually indicate good quality for bone development. A good length and width of the bird are clear indicators for high vigour, longevity, and provide enough space for development and growth of various internal organs. Birds with tapered or narrow backs do not have ample space for maximum body meat development.

Feeling flatness of back.

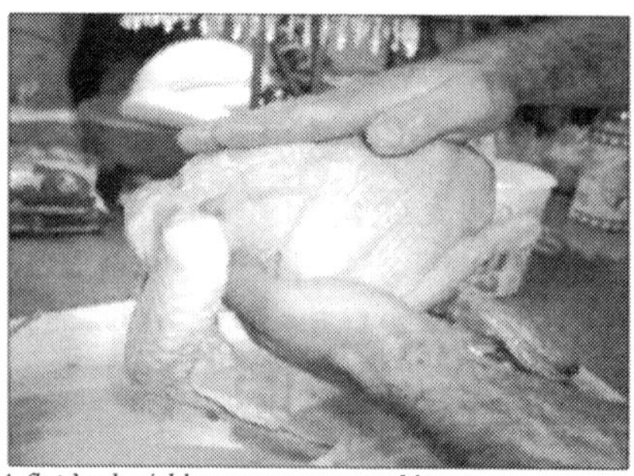

A flat back yields a more presentable carcass.

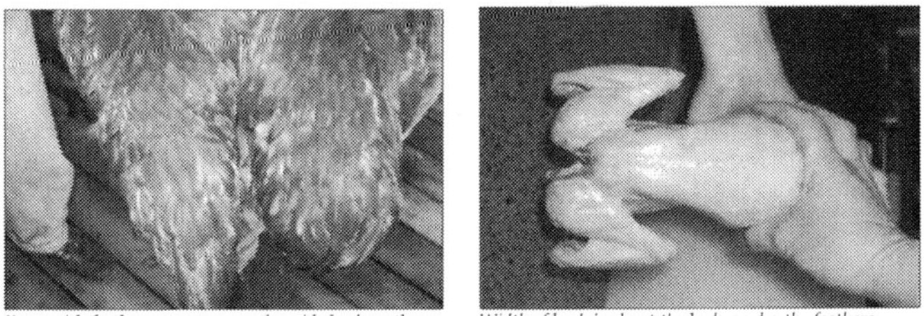

Even with feathers, you can see the wide back on the right and narrow back on the left.

Width of back is about the body under the feathers. Hand shows position to appraise.

Figure 6.4: The back of a good broiler bird should be flat, long, and wide.

Source: *https://www.slideshare.net/PakRose1/poultry-breeding*

The keel of the birds should be assessed for length and straightness while the breast should be examined for development of good meat proportions. More customers are usually appealed by the breast meat of the broiler birds. When assessing the flesh around the breast area, the birds should be positioned in an inverted position.

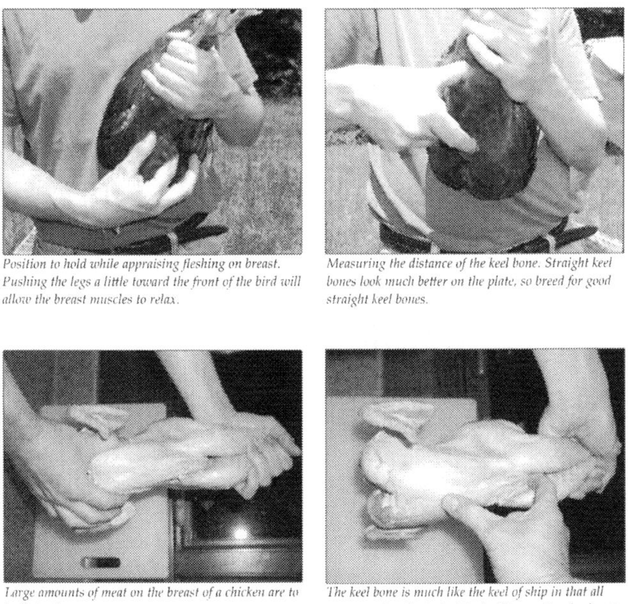

Position to hold while appraising fleshing on breast. Pushing the legs a little toward the front of the bird will allow the breast muscles to relax.

Measuring the distance of the keel bone. Straight keel bones look much better on the plate, so breed for good straight keel bones.

Large amounts of meat on the breast of a chicken are to be desired.

The keel bone is much like the keel of ship in that all organs rest on it. A short keel will cause the checken to look meaty, but leaves little room for healthy organs.

Figure 6.5: Broiler chicken selection based on the length and straightness of the keel.

Source: *https://www.slideshare.net/PakRose1/poultry-breeding*

To determine the general growth rate of the birds, their body weight should be obtained. It is advantageous to actually use body weight of the birds as an impartial record that helps in comparing the birds in different generations. Studies have shown that a male broiler bird should attain a body weight of about five pounds while the female should have about three pounds five pounds by the age of 16 weeks.

Appetite of the birds can also be use when selecting broiler birds for breeding or business purposes. The body of the birds usually grow based on the amount of feeds consumed. However, the conversion rate of these feeds to body size and weight gained should also be high. It is important therefore to always consider those birds showing strong appetite for business and breeding purposes.

Table 6.2: The different meat type breeds that are commonly used by farmers for commercial and backyard meat production

	**Cornish Cross** This is one of the best breed reared for the purposes of meat production. Males are able to reach about twelve pounds within a period of 6 to 8 weeks. Because of its fast growth rate, most commercial broiler farmers prefer rearing the breed for meat productions. Other than fast growth rate, this breed has the best tender and tasty meat compared to other common meat type breeds. This breed has large thighs, broad breast, and large legs with enriched yellow skin that make the breed perfect to be used as table meat. It is heavy feeder; harvest time is around 4–6 weeks, and has the best white meat. The breed is only white coloured
	**Jersey Giants** This is US origin meat type breed developed with an aim of replacing turkeys in meat production. There are purely bred and their average weight range from about 12 to 14 pounds. Their growth rate is a bit lower as compared to other commercially reared breeds. For this reason therefore, they are not often used for commercial meat production. They are the best meat type breeds for backyard chicken farming.

	**Bresse** These are very popular meat type breeds found almost everywhere in the world. They are usually quite expensive and have feet that are bright blue in colour. They are also considered as the best in providing the tastiest meats in the whole world. Their meat is also tender. A male Bresse broiler chicken can weigh up to seven pounds while the females weigh around five pound by the time they reach market age. Most of them reach maturity at 16 to 20 weeks of age. This breed has the best tender and tasty meat and is also docile. However they do not have meaty bodies. There are different varieties including blue, white, grey and black coloured.
	**Orpington** This is another heavy meat type breed with a capacity to grow up to seven to eight pounds for the females and ten to twelve pounds for the male chickens. They are good egg layers too, but they are mostly preferred because of the tender and amazing flavour of their meats. This breed has a broad and curvy-shaped body with a short back that appear U-shaped. Their growth rate is usually slower compared to other commercially reared broiler. They occur in different colour such as white, blue, black, and buff.
	**Freedom rangers** These are one of the best pasture-fed meat type breeds. They were developed for free-range meat production. They are able to flourish very well with poultry feeds that lack adequate protein and are the best at scouting for food on the field. This actually makes freedom rangers the best meat type chickens to scavenge for food around in large pen. They are also the perfect and versatile feeds consumers. However, they take a long period time to reach their market live-weight, compared to other meat type breeds such as Bresse and Cornish broiler chickens. A fully matured male bird can weigh up to 6 pounds while female weigh 5 pounds. Harvesting can be done after about 9 to 11 weeks of age. They occur in different colour such as black, tricolour, and red.

Source: *http://ouroneacrefarm.com/guide-choosing-chicken-breeds-pick-best-breeds-flock/*

BROILER POULTRY HOUSING

Broiler birds are mostly kept purposely for the production of meat. They are modified to enhance their growth rate and reach maturity within the shortest time possible. Proper management and housing is a priority for farmers to achieve maximum meat production from their broilers. Studies have shown that broiler chickens are more reliant on environmental conditions than other poultry birds. Therefore, for maximum production to be achieved, proper housing for the broiler bird should be emphasized.

Figure 6.6: A typical broiler house.

Source: *http://www.roysfarm.com/broiler-poultry-housing/*

When constructing the house for the broiler chickens, there are different materials that one can use such as concrete, wood, or even steel supports and trusses. From side to side of the wall, the house should be clear distance. To support the weight of the roof, farmers may use concrete poles or steel trusses. It is important to always ensure that the birds are provided with suitable beddings (litter) on the floor of the house. Some of the litter materials that can be used in the house of broilers include wood shaving, rice hulls, peanut hulls, and sand. The dimension of the house for broiler birds may be determined by the number of birds expected and the environmental conditions surrounding the place. The best dimensions for commercial broiler rearing, the house should be about forty to fifty feet wide and four hundred to six hundred feet long. The side wall may be constructed about eight feet high. This broiler housing dimension should not be fixed; it should vary depending on the number of the broiler birds reared and other purposes. For the purpose of fresh air circulation in and out of the broiler house, dropped ceiling is highly recommended especially is areas with hot climatic conditions.

Housing

There are different types of broiler housing systems. Broiler birds can be housed in wire floor, slatted floor, cages, or in a deep-litter house. Due to space limitation, wired floor, slatted floor, and cage rearing are rarely used because of various accompanying issues such as leg weakness, blistering on the feet, and high initial costs for investments. The most common housing systems for broiler birds include the all-in all-out system, also called single batch at a time and multiple batch rearing system.

Single batch system-the farmer rears one batch at a time for a specific period of time. In other words, chicks of the same age are purchased or hatched and reared until they reach market weight. They are marketed in a single slot. The important of using this broiler housing system is because it enhances proper hygiene and proper management of the broilers, prevention and control of various sub-clinical diseases is also made easier using this system, the spread of infection is usually minimized thereby lessening the number of birds mortality in the house, lastly but not least this system enhances efficiency in feeding and fast growth rate for the broiler birds. However, this housing system is only useful in small scale broiler farming. The other disadvantage of using this system of housing is that fixed and working capital is usually high per bird.

Multiple batch housing system-this system involves rearing different broiler batches at the same period of time. The interval between one batch to the other is usually about 1 to 4 weeks. In other word, the farmer have a schedule that ensure that new one-day-old chicks are bought and the mature chicken sold on a monthly, every 3 weeks, fortnightly, or on a weekly basis. Broiler chicks are raised for a maximum of 5–6 weeks of age or until they attain the required body weight for table meat.

Broiler Floor Space, Feeder Space, and Water Space

There are several factors that determine the required spacing for the broilers. These include the body weight of the birds as they reach marketing period, marketing age of the birds, the type of housing system used for the broilers, and the ambient temperature. The spacing of the water dispenser and feeder is also determined by various health conditions and environmental conditions surrounding the birds in the house.

Table 6.3: The approximate spacing of the floor, water dispenser, and feeder required for the broiler birds.

Age	Floor space/ bird	Feeder space/ bird	Watere space / bird
Up to 18 days	450 cm2 (0.5 sq.ft.)	3 cm	1.5 cm
From 19 days to 42 days	1000 cm2 (1.1 sq.ft.)	6-7 cm	3 cm

Source: *http://agriculturenigeria.com/farming-production/livestock/poultry*

Heating

Proper temperature maintenance in broiler house is of paramount in productivity of the birds. The growth and productivity of the birds highly depend on how farmers are able to maintain house temperature. Providing a suitable source of heat especially during extreme cold winter seasons is highly recommended. This is because most broiler chickens are not able to maintain their body temperature until they are 2 weeks old. During the first 2 weeks of age, farmers are advised to maintain the temperature in the house of chicks at about 85 to 90 °F without any significant variation. There are several heating systems that can be used in broiler houses. These include radiant tube heaters, force air furnaces, pancake brooders, and radiant brooders among other heating systems.

Brooders and tube are mainly used in floor heating, while the hot air furnaces are for heating the air which then warms the entire house. When using hot air furnace it is important to always remember the principle that says "hot air is lighter than cold air" this causes the hot air to always be at the ceiling of the house while the cold air to be on the floor of the house. This condition can actually lead to stratification. It is important therefore to always use circulation fans in mixing the warm air at the top with cold air at the bottom of the house. When mixing hot and cold air, the hot air usually helps in drying up the litters in the house. Paddle fans can also be used in mixing the air, but extra care should be taken not to expose the chicks to the strong drafts from the fans.

Ventilation

Broilers require proper ventilation system for delivering fresh and clean air in the house. Proper circulation is also important in removing excess moisture, heat, and other unwanted gases from broiler house. There are several types of proper ventilation of the broiler house. These include temperature controller and thermostats, fans, evaporative cooling system, and air inlets among others. Ventilation in broiler house can be classified into two main system. These are the hot weather ventilation system and cold weather ventilation system.

Figure 6.7: Hot weather ventilation system.

Source: *http://thepoultryguide.com/poultry-ventilation/*

Hot weather ventilation system: this system is used during the hot weather season. The aim of this system is to cool the birds. One of the best type of cooling is the tunnel ventilation, which consists of very huge inlets for cold and fresh air usually at one end of the house and a fan is usually erected on the other side in order to help in circulating the air all-round the house. The fan is also used to draw the air in to the house at a velocity of about five hundred feet per minute. The aim of using this system is to help in the removal of the excess heat from broiler house thus causing a cooling effect in the whole house. Using tunnel ventilation system is not adequate.

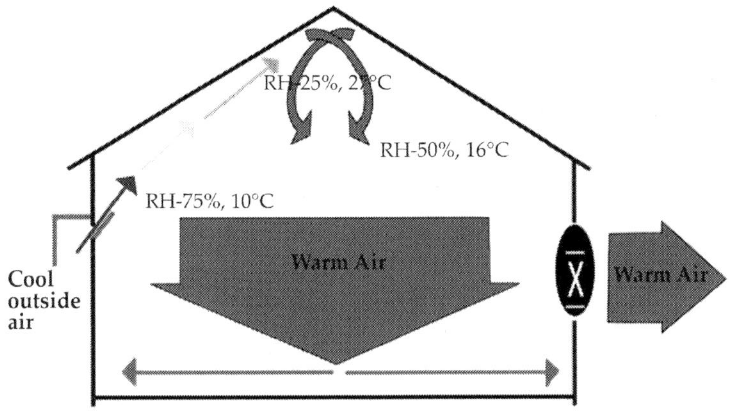

Figure 6.8: Cold weather ventilation system.

Source: *http://simsroofingandsiding.com/proper-attic-insulation-and-ventilation-extends-roofing-life/*

Cold weather ventilation system: this involves the provision of fresh and cleans air in broilers house using negative pressure ventilation mechanism during cold weather. The aim of this system is to minimize heat loss through removal of excess moisture in the broiler house. Fans are used in exhausting air out from the house through the creation of a negative pressure in the house. Reduction of pressure in the house actually causes entrance of fresh and clean air into the house through various planned air inlets. Installation of these air inlets is usually done on the high side of the house walls or in the ceiling. The reason why air inlets are installed high in the ceiling or on the walls of the house is to facilitate air circulation across the ceiling areas and mixing this air with the warm air that is already there. Once this fresh air is warmed up, it is then drifted around using fans to the rest of the house for warming purposes. Farmers are advised to regularly check the levels of the temperature in the broiler house in order to determine necessity of heating or cooling. This actually helps in minimizing the costs incurred for cooling or warming up the house. For the extensive commercial broiler production, farmers are encouraged to always install automatic or computerized heat regulators in the houses.

Lighting

This is one of the essential factor for commercial broiler production. Broiler breeding and processing companies have set specifications for the intensity of light in brooding and growing stages of the birds. During brooding stage of the birds, the recommended light schedule should be 24 hours. For the growing broilers, the required lighting should be 23 hours of light and 1 hour of darkness until they reach the marketing period. The 1 hour is specifically included in order to acclimatize the birds to darkness especially in case of power outage, which may result to trampling and panic of the birds. The recommended colour of the light should be a dimmable white light.

FACTORS AFFECTING BROILER PERFORMANCE

To evaluate and determine the performance of broiler birds in the farm, there are several measures that breeders can actually use. Some of these measurements include the liveability, growth rate of the broiler chickens, morality rate of the broiler chickens, days to market, and feed efficiency. One of the most costly expenses in the production of broiler birds is the feed costs. Therefore, broiler farmers use efficiency of feeds in their birds as their primary tool for evaluating performance of their flocks. There are several ways in which feed efficiency in broilers can be obtained. One way of doing this is by dividing the total feeds consumed by the total weight gained. The resulting value should be around one point eight for about forty two days old broiler chicken. The lower the value obtained (known as Feed Conversion Ratio-FCR) the more efficient the broilers are in the use of feeds consumed. Some places like in Europe, feed efficiency is obtained by dividing the gained weight by total consumed feeds. The corresponding figure should be around zero point five. The higher the value, the more the efficient the birds are in conversion of feeds. The following are various examples used in measuring performance of broilers.

- The normal value obtained in this calculation is usually about 97 to about 98%
- The value obtained should be one point eight or less for a period of 6 weeks of age
- The higher the value obtained the best performance index of the broiler birds. A value of hundred or more is usually preferred to indicate better broiler performance.

A broiler farm economy index value of about two and above is usually a perfect indicator for good broiler business management and optimal broiler birds' performance. If the value is less than one point three, it indicates that the broiler birds are poorly managed.

Currently, more broiler farmers have shifted from standardized broiler growing programs to those programs that are focussed towards meeting specific economic conditions and various local goals for broiler production. Numerous elements influence both the consumption of feeds and growth rate of the broilers, thus affecting their feed efficiency. One of the most important factors influencing performance of broilers is the level of energy in feeds. Previously, feeds with high level of energy were given to the birds-for instance, the starter feeds were 3,000 kcal/kg (about 1361 kcal/lb.) while the finishers were given up to 3,200–3,300 kcal/kg (translating to 1,452–1,497 kcal/lb.). Because of the high cost of production in broilers and other associated management problems, poultry feed ingredients with much lower levels of energy are being used today in formulation and production of broiler feeds. Additionally, broiler chickens are nowadays being grown in different and wide variety market body weight. Efficiency in feed conversion decreases as the birds becomes more and more old. The following are some of the major factors that influence performance of broiler chicken before attaining the required market weight.

House temperature-ambient temperature in broiler house is one of the main non-dietary factors that affect broiler performance through their feed conversion efficiency. Broilers are considered to be warm-blooded animal (homeotherms), hence they are able to maintain their body temperature without concentrating on other environmental conditions. Studies have shown that performance of broiler chicken is usually at optimal when there is less or no variation on the house temperature for a period of twenty four hours. For this reason therefore, it is highly recommended to always supplement the natural temperature source with other artificial source of heat in broilers house, especially during extreme cold winter seasons.

It is has been revealed that broiler birds have a tendency of consuming more feeds during thc cold weather seasons. Most of the calories that they get from consumption of the feeds are used in maintaining their normal body temperature. This usually results to poor performance among the birds because more calories are used for sustaining body temperature rather than been converted to meat. Broiler farmers are therefore advised to always provide artificial heating source during cold weather and also at night when the temperature falls. It is also important to regulate environment temperature. This is because high environmental temperature result to less consumption of feed thus less efficiency in feed conversion. It is also important to remember that broiler birds also require energy for various biological mechanisms used in cooling such as panting during hot seasons just as they use energy to warm themselves during the cold weathers.

Quality of House Litter

The quality of litter used in the house of broilers is another factor that influences the performance of the birds. This ultimately influence the profits obtained from the birds at the market level. House litter can be described as the combination different material in the house of the birds such as bedding materials, feathers, excreta, wasted water, and wasted feeds among other materials. For maximum performance on broilers, proper management on house litter should be done.

Feeds wastage and feed deprivation -broiler farmers are advised not to place excess feeds in feeding troughs of the young chicks. This helps in minimizing wastage of feeds that may result to lower feed conversion.

Automated poultry feeders can be used to reducing feeds wastage through adding small quantities of feeds frequently over short period intervals of time. This not only does it help in preventing feeds wastage, but also encouraging the young and growing broiler chicks to eat quickly and more from the automatic feeding equipment.

Figure 6.9: Automated poultry feeders.

Source: *http://www.turkishbusinessplatform.com/ar/product/5438-automatic-poultry-house-equipment*

The deprivation of feeds can be a main factor that impact negatively on the performance of growing broiler birds. It contributes to lesser feed conversion. Feeds deprivation can be as result of introducing automatic feeding system in broiler house for the first time. It is therefore advisable not to raise the feeders too early or too high during production cycle of the birds. Some of the common effects of early deprivation of feeds include uneven broiler growth and poor uniformity.

Infections and Culling

Feed conversion in broilers is greatly affected by the general health status of the birds. Performance of sick broilers is usually poor. Farmers are encouraged to regularly check for various disease signs and symptoms, and if any, proper and immediate actions to control the disease should be taken. Proper handling of vaccines and other medications is recommended because poor handling may cause severe reactions in the bodies of the broiler chickens thus lowering feed conversion and body weight gain. Culling should be done as early as possible for the severely diseased birds with lesser chances to reach to market.

Generally, broilers with poor health are likely to convert feed poorly. This is because their appetite is usually affected negatively and more feeds are assimilated for maintenance purposes. Heavy infestations of both internal and external parasites also affect conversion of feeds to body weight. Internal parasites and microbes may cause severe damages on the intestine thus resulting to poor digestion and reduced nutrients absorption. Broilers chickens suffering from coccidiosis are likely to have poor nutrients absorption thus less feed conversions efficiency. This is because of damages caused by the oocytes on various cell linings along the gut. There is also another health related condition referred to as "feed passage" that is commonly seen in broilers. This involves presence of undigested feed materials in the excreta, which consequently affects efficiency of feed conversion. Although the exact causes of this condition are not known, it is believed that it may be as a result of microbial challenges in the digestive tract.

Figure 6.10: Feed passage in poultry.

Source: *https://www.slideshare.net/DSM_Animal_Nutrition/the-role-of-intestinal-health-in-broiler-production-e-o-oviedo-at-espn-2013*

Human Factors

Researches have shown that there is a strong relation between broiler fear levels caused by various human behaviours towards the productivity and performance of the broiler birds. To enhance production and performance of broilers, it is important to lower the level of fear caused in broilers chickens. Previous studies have indicated that the mortality of broiler chickens is usually recorded during the first week of their age. High movement speed usually as a result of high fear levels caused by human beings is one of the main causes of high mortality. In other words, this means that young broiler chickens are usually more susceptible to different human stressors such as movement speed of the stocking personnel. However, this phenomenon can actually be lessened through broiler chickens adaptability to various human behaviours as they mature.

WELFARE ISSUES IN BROILER PRODUCTION

Due technological advancements experienced in modern agricultural industry, different methods have being adopted in breeding and production of broilers. Artificial selection of the bird has actually led to development of broiler chickens that grow and develop to slaughter weight within a short period of time than normal poultry meat type birds. This has led to various welfare issues on current broiler breeds. This is a huge concern not only in the local market, but also at international market. For instance, in the year 2011, the US broiler chicken production was estimated to about 9 billions. More health-related welfare has increased with increased in broiler production.

Cardiovascular Dysfunctions

Most current broiler producing companies are using artificial selection and birds' husbandry to enhance the growth rate of the broiler chickens. This has hereditarily induced various mismatches between various internal organs supplying energy and those organs that consume the energy. Due to the rapid growth rate in broiler chickens, most of them usually develop various disorders related to metabolic failure such as ascites and sudden death syndrome (SDS).

Sudden death syndrome is a poultry disease related to the failure of the heart that mainly affects the fast growing male broiler chickens. Some of the main clinical signs and symptoms of this condition include flapping wings, imbalance, most male broiler are seen crying out and falling on their sides or back. Death usually occurs within a few minutes. United

Kingdom recorded about 0.8% of SDS incidences in the year 1993. In the year 2000, SDS related death rate ranged from 0.5% to about 3% in most of European countries.

Ascites on the other hand is usually characterised by various clinical signs such as heart dilation, hypertrophy of heart, liver dysfunctions, pulmonary dysfunctions, fluid accumulation in the abdominal cavity, and hypoxemia among other signs. Occurrence and development of ascites in broiler chickens is usually gradual and the birds can suffer for a long period of time without noticing any signs until it die. More than nineteen million broilers usually die due to ascites and other heart failure related problems in UK every year.

Skeletal Dysfunction

Most broiler chickens are bred with an aim to increase their muscles and body weight within the shortest time possible. Their breasts are usually wide and full of meat compared to other indigenous meat type birds. This actually shifts the centre of gravity of their bodies and this seriously affects the way they walk. Most birds develop legs and ribs related problems due to excess body weight. Some of the commonly reported skeletal complications in broiler chickens include necrosis of the femoral head, dyschondroplasia, deformities of varus and valgus, and osteodystrophy. These conditions usually affect the movement of birds and most of them end become lame thus spending most of the time sleeping and lying on the ground. Due to lack of movement, most broiler chickens may develop bone ossification resulting to abnormalities of the skeletal.

Integument Lesions

Most broiler chickens usually develop lying and sitting behaviour due to rapid growth of their bodies. About 75% of sitting and lying behaviours is usually reported in broiler chicken within the first seven days of their age. By the time the birds are about 35 days of age, about 90% suffer from sitting and lying behaviours. Most of these behaviours are related to dermatitis resulting from accumulation of ammonia in house litter. Most issues of dermatitis in broilers is usually characterised by epidermis necrosis and hyperkeratosis on the affected areas. Some of the clinical signs seen on the broiler due to dermatitis include lesions on their foot pad, blisters on the breast, and burns on the hock.

Stocking Density

Most broiler chicken is usually reared in large number in one house. Stocking densities vary from one country to the other. In UK, stocking density range from 25 kg/m^2 to about 45 kg/m^2 (this translate to about 12 to 25 broiler birds in every square metre). If stocking density of broiler chicken reared under a deep litter house exceeds 30 kg/m^2, the growth rate of the birds and feed intake is usually reduced. This is caused by reduced space in the house which limits heat loss produced by the birds through metabolism in their bodies. Some of the common broiler health problems related to high stocking densities includes soiled plumage, blistered breast, and lesions on the foot pads.

Ocular Dysfunction

Most broilers chickens are housed under various lighting conditions to enhance and maintain a fast growth rate. Some of these conditions include continuous darkness, continuous light (incandescent and fluorescent) or under dim light. Most chicken reared under this lighting condition develop various eye related abnormalities such as enlargement of ocular, avian glaucoma, shallow anterior chambers, and macrophthalmos among other conditions.

CHAPTER 7

LOCAL AND INTERNATIONAL POULTRY MEAT PRODUCTION INDUSTRY

INTRODUCTION

This chapter examines breeding and production of poultry in the United States (US) and other countries based on the trends in the production and trading, challenges in trading poultry products, and various main markets around the world. This covers all the poultry birds raised for meat production. The industry of poultry meat production involves broiler production (chickens) and turkey meat production. There are other smaller poultry meat producing industries including industries producing guinea fowl meat, geese meat, and duck meat. However the primary focus in this chapter is on the production of turkey meat and broiler (chicken) meat. The US is the largest producer of poultry meat both at local and international levels.

Based on previous research, it has been revealed that the US accounts for the largest poultry meat production worldwide. Between the years 2006 and 2012, the total production of poultry meat in the US was approximated at one-quarter of the total poultry meat production worldwide. During this period, the total poultry meat export to other countries increased to about 40%, which accounted for more than 35% of the total turkey and broiler exports in 2012.

The US industry of poultry meat production is mainly vertically integrated with an increasing efficiency because of various factors such as technological advancements, wide research and development strategies, and economies of scale. However, the poultry meat production also has various challenges such as high prices of the poultry feeds and lower demand of poultry meat in the local market. The US faces also a huge competition in the exportation of the poultry meat products to other countries. Some of the close poultry meat producing competitors include Brazil, China, Thailand, and the European Union, which are currently increasing their production and exports.

The total value of poultry meat production in the year 2012 was estimated at around 30.2 billion US dollars. Chickens are the most preferred poultry meat in the US region and is consumed all year round. Turkey meat is the fourth most consumed meat in the US. The main reason for this poultry meat preference in the US is its competitive pricing structure, its medical value being considered healthier than other red meats, and its relatively shorter preparation and cooking times. Additionally, there are no prohibitions on poultry meat consumption by groups and members of certain religions.

Chicken meat (broiler) is usually sold in various forms after processing, including frozen, chilled, and fresh. It is actually processed and packed in different forms such as whole chickens, chicken legs, chicken breast meat, and chicken wings. On the other hand, turkey meat is also processed and sold primarily as whole dressed turkeys. It is sold in different forms such as frozen, chilled, and fresh. Poultry meat is also converted into other ready-to-cook products such as sausages, deli meats, and other microwaveable meals.

LOCAL POULTRY MEAT INDUSTRY AND INTERNATIONAL MARKET

The US has the best poultry production structures that have enabled the region to maintain a high production of various poultry products both in the local and international markets. Indeed, the US has been considered one of the best and largest poultry producers worldwide for the past few decades. Hatching eggs are usually provided by the breeding companies to various poultry producers that have being integrated into the industry vertically, commonly referred to as 'integrators'. Subsequently, these poultry producers offer breeding contracts to various independent 'grow out' farmers to breed and raise live birds. Once the birds have attained the required live-market weight, they are returned to the integrators for slaughtering and processing purposes into either cuts or whole dressed birds. Integrators and other individuals may decide to process the poultry meat further into other finished poultry products sold within and outside the country (local and international market). The following figure illustrates the structure of poultry industry in the US, which is made up of various poultry producers, principal poultry inputs, end or finished poultry products, and key markets.

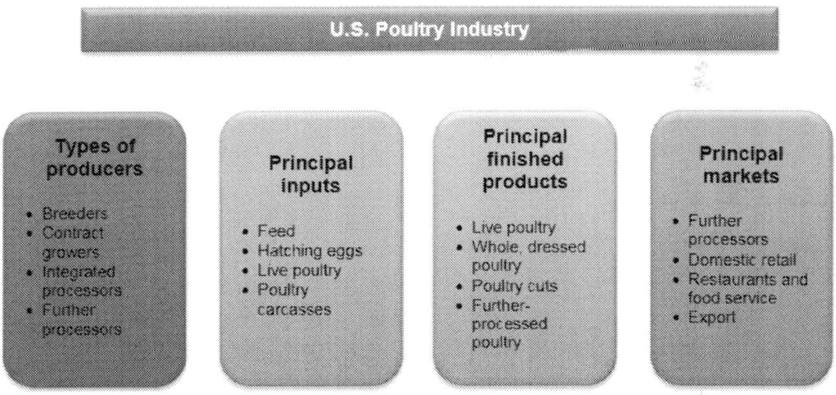

Figure 7.1: The structure of poultry industry in the US.

Source: *https://www.ers.usda.gov/data-products/chart-gallery/gallery/?page=39&topicId=0*

The total poultry meat production (broilers and turkey) was estimated at about 25.6 million metric tonnes between the 2006 and 2012 in the US. From 2008 to 2009, there was an economic recession in the country, which led to a reduction in the total poultry meat production, resulting in a huge decline in the demand of poultry meat. During this period, the average broiler meat produced by the top three integrators accounted for almost 50% of the total chicken meat production in the nation. On the other hand, the top three turkey meat producers accounted for over 50% of the total poultry meat produced in the nation. Some of the largest broiler meat producing states in the US included Mississippi, Alabama, North Carolina, Arkansas, and Georgia. These states accounted for about

60% of the total broiler meat produced from 2006 to 2012. For the turkey meat production, both North Carolina and Minnesota were the largest producers and accounted for 31% of the total turkey meat produced from 2006 to 2012.

Global production on both turkey meat and chicken meat from 2006 to 2012 rose from 7.5 million metric tonnes to 8.3 metric tonnes, which translates to about a 25% increase. The US is actually the largest poultry meat producer accounting for about one-quarter of the total world production. China and Brazil are the second largest poultry producers accounting for 15% each of the total world production. The third largest poultry meat producer is the European Union accounting for about 13% of the total global production. Among these four top poultry producers, Brazil accounted for the highest growth rate (of about 36%) in 2006 to 2012. This was followed closely by China with about a 32% growth rate over the same period. These growth rates were attributed to the expansion of capabilities on domestic poultry production and modernization of the poultry production.

Studies also showed that the total consumption of poultry meat between 2006 and 2012 rose from 70.3 metric tonnes to 86.4 metric tonnes, which translates to about 23%. The US was followed closely by China, the European Union, and Brazil who are also the biggest consumers of poultry meats, all accounting for 60% of the total world poultry meat consumption.

The international export of poultry meat rose from 7.1 million metric tonnes in 2006 to around 10.8 million metric tonnes in 2012, which translated to an approximate 52% increase. Statistically, it has been revealed that the largest poultry meat producers are also the main poultry meat exporters. The US has been leading in the exportation of poultry meat for a long period of time. Significant increment in the poultry meat exportation has also been observed in Brazil, especially for the last few decades. Both the US and Brazil accounted for about three-quarters of the total exportation of poultry meat worldwide from 2006 to 2012. Between 2006 and 2012, poultry meat exports of the US increased by around 39%, while that of Brazil increased by about 38%.

Global poultry meat importation rose by about 32%, from 6.8 million metric tonnes to about nine million metric tonnes from 2006 to 2012. Most of the importations have always been lower compared to the exportations.

The five largest poultry meat importers include the European Union, Mexico, Japan, Russia, and Saudi Arabia, all accounting for about 55% of the total importation worldwide in the 2006 and 43% in 2012. The huge decline in the global importation was attributed to the reduction of poultry meat importation by about 55% in Russia between 2006 and 2012, with the aim of becoming more independent in the production of poultry meat.

POULTRY MEAT PRODUCTION ISSUES

Poultry meat production in the US and other international breeding states face numerous issues that affect their competitiveness in the global poultry meat production. Among the noticeable issues include exorbitant pricing on poultry feeds, which lower profitability of poultry meat production, and barriers on the foreign trades, which reduce the exportation of poultry meat to other countries such as China and Russia.

Poultry feeds are one of the priciest poultry inputs in the production of various poultry products such as eggs and meat. Based on various previous agricultural studies, poultry feeds account for about 65 to 75% of the total cost in poultry production. The main ingredients used in the production of poultry feeds are soybeans and corn, which had increasing prices on a yearly basis. Poultry feeds are mainly influenced by several factors (both long term and short term) that impact the supply and demand of various poultry products. Internationally, supplies of poultry products are affected primarily by various factors such as controls on exportation of poultry products and environmental conditions such as drought. On the other hand, the demand for more poultry product at international levels is hugely impacted by things such as increase in personal income, increase in population, and expansion on biofuel production among other factors. However, due to the reduced demand of poultry products among various domestic customers and high costs on poultry feeds, the poultry industry in the US and other international producers has been deteriorating consistently since the 2006 to date. To help in enhancing the efficiency of the poultry industry, the integrators have terminated various facilities that are underperforming with an aim to reduce the production costs and enhance the profitability of various productive facilities.

Although the US is still a highly competitive poultry meat supplier both in domestic and international markets, poultry meat exportation to various key international markets is disrupted, mainly by various trade-related measures legislated by China and Russia in 2010. For instance, China enacted countervailing and antidumping duties (AD/CV or AD/CVD duties) on some of the poultry meat products produced in the US. This, according to the World Trade Organization (WTO) viewpoints of August 2013, violated several obligations enshrined in the WTO Antidumping Agreement and Subsidies Acts. Russia on the other hand enacted legislations prohibiting the importation of poultry products that are treated using chlorinated rinse (a commonly used pathogen reduction treatment in the US). These regulations greatly reduced the exportation of US poultry products to China and Russia from 1.6 million metric tonnes in 2008 to as low as 0.35 metric tonnes in 2011, which translated to about 79% decline. Russia and China were the two main and largest importers of various poultry products from the US from 2006 to 2009, which on average received around 34% of US exports annually. However, importation of US poultry products by these countries declined between 2011 and 2014 by about 11%.

THE STRUCTURE OF POULTRY MEAT INDUSTRY

Number and Consolidation of Poultry Firms

Consolidation of poultry breeding and production industries is one of the main factors that have facilitated the growth and competitiveness of various key poultry producers in the world. Consolidation has actually ensured that there are only a few, large and more efficient poultry producing and processing facilities. Between 2002 and 2007, the total number of poultry producing farms that recorded sales of live poultry birds for meat production decreased marginally due to a lower number of farms producing broiler chickens and turkeys. The biggest decrease (15%) was reported in broiler and other meat type chicken rearing farms. This actually replicates the consolidation of various small poultry farms into fewer, larger, but more efficient poultry farms. It is a strategy that has often being used in various successful poultry meat producing countries in the world based on economy of scale. Farms tailored towards rearing chickens for egg production (pullets and layers) and other poultry birds such as geese, guinea fowls, quails and ducks indicated a 9% and 11%, respectively.

Table 7.1: US poultry farms by type, 2002 and 2007

Type	2002	2007	Percent change
Layers and pullets	26,814	29,140	9
Broilers and other meat-type chickens	32,006	27,091	(15)
Turkeys	8,436	8,284	(2)
Ducks, geese and other poultry	19,732	21,857	11
Total	86,988	86.372	(1)

Source: *https://www.ers.usda.gov/data-products/chart-gallery/gallery/?page =39&topicId=0*

In the last few decades, the consolidation of the poultry industry has been slowing down considerably. Between 2006 and 2012, the total production of the three key broiler producing companies declined by 10%, while the total share of the top 10 producing companies maintained their production at about 77%. The biggest decline in poultry industry consolidation happened between 2008 and 2009, mainly because of a decline of 12% in Pride Corp Pilgrim production. This was actually coupled by declines in poultry production in other large producers such as Perdue Farms Inc. (which accounted for a 2% decline) and Tyson's Food Inc. (which accounted for about 1% decline). These declines in poultry meat production were attributed to economic recession that occurred in 2008 to 2009, which negatively affected the demand and consumption of poultry products in the US.

However, there were some consolidations of poultry producing industry from 2006 to 2010. These included the purchase of Gold Kist Inc. by the Pilgrim's Pride; the sale of various turkey facilities of Pilgrim's Pride to other turkey producing companies in the US, and the purchase of various broilers producing facilities belonging to Peterson Farms by

Simmons Food Inc., among other consolidating strategies. Consolidation of various broiler producing industries has being happening up to today and has really helped in enhancing poultry meat production industry in one way or the other. There was a slight decline (5%) on the production of the top three turkey integrators from 2006 to 2012. However, consolidation of the top 10 turkey producing firms slightly rose by 3% over the same period of time.

Table 7.2: Share of US production accounted by 3, 5, and 10 firms

Sodom	2006	2007	2008	2009	2010	2011	2012
Broilers:							
Top 3 firms	53,1	50,9	49 3	46,5	45,7	46 6	47,6
Top 5 firms	62,6	60,9	60 5	57,5	57,0	58 7	60,7
Top 10 firms	76,6	75,8	75 7	74,8	74,5	76 7	78,8
Turkeys:							
Top 3 firms	52,0	50,7	50,5	53,1	51,2	50,2	49,5
Top 5 firms	58,8	58,1	58,1	60,9	61,0	59,7	59,5
Top 10 frets	75,2	75,0	74,7	78,9	78,9	77,6	77,5

Source: *https://www.ers.usda.gov/data-products/chart-gallery/gallery/?page =39&topicId=0*

Vertical and Horizontal Integration

For the past few decades, production of broiler and turkey breeding has being highly integrated vertically. This has enabled production of more than a 90% production of broilers and turkeys through vertical integration by US companies. Most of these vertically integrated companies own some of the biggest hatcheries; poultry feed millers, slaughter houses, and processing companies. Most of the operation involved in rearing the broiler and turkeys for meat production is not done by the integrators; rather they usually subcontract the operation business to various farmers to raise the chicks until they attain the required live-market weight. Vertical poultry breeding integration is one of the best modern breeding techniques that enables most poultry processors to uphold high level of control over their poultry products. This helps in production of the best quality and standardized tasty meat with an aim to maintain a controllable and steady poultry supply of the birds for processing. Vertical integration also helps those countries that deal with high level of poultry production like the US to take advantage of economies of scale.

Most poultry companies in the US and other high producer countries concentrate mainly on poultry products. However, some poultry producing companies have integrated horizontally with several other agricultural companies, such as those dealing with meat packaging business. For instance, apart from production of poultry, some companies like Kraft Food Inc., Tyson's, and Cargill Inc. are also involved in production of other agricultural food products. Additionally, there are other poultry producing companies that act as subsidiaries of other food producing companies. For instance, Carroll's Turkey LLC is a small company that deals with turkey production, but it is own by another company called Smithfield Food Inc. which mainly deal with production of pork.

Globalization of Poultry Industry

Poultry breeding sector is own and run by large global organisations. One of the largest global companies involved in breeding and production of poultry is Aviagen Group Ltd. This is a German poultry meat company operating in different countries in the world, with the US being included. Aviagen Group Ltd is a combination of two companies referred to as Aviagen Group (for broiler chicken) and Aviagen Turkeys Inc. (previously referred to as Nicholas Turkey Breeding Farms). Another poultry producing companies with a wide worldwide network is known as Cobb-Vantress Inc., which has other subsidiary companies worldwide that help in distribution of various poultry products produced by the company.

Generally, poultry production industry in the US is actually not vastly globalized at the integrator level. This is because most of the poultry companies in the US are locally owned and ran. However, some companies like Tyson's, which is the largest integrated poultry producer in the US, own other companies directly or through joint venture in other several countries such as Mexico, India, Brazil, and China. Cargill VAM is a US owned poultry producing company that specializes in the production of turkeys, but it is a subsidiary and privately run and owned by Cargill Inc., which is a large poultry producing company with modern poultry rearing facilities in several countries like France, Brazil,

Ireland, Canada, Honduras, Thailand, Netherlands, United Kingdom, and Nicaragua. There are also other non-agricultural companies that have also taken part in production of poultry oversees. For instance Goldman Sachs, which primarily practices on financial services has invested in ten different poultry producing companies in China.

PRODUCTION PROCESS AND MARKETING CHANNELS IN POULTRY MEAT

Production and marketing of poultry products in the US and other leading producers is one of the most significant elements that drives the growth and development of the industry. There are six key stages through which conventional poultry production takes place in the US and other poultry producers. These stages, which are shown in the Fig. 7.2, are similar in the production of turkeys and broilers. Integrators are usually the primary entities controlling the entire production process of poultry breeding. The main difference that occurs between turkey integrators and broiler integrators is that some turkey integrators actually breed and rear their poults (turkey offspring) in-house, while very few chickens or broiler integrators actually own grow-out poultry operations. Some poultry meat integrators may own other poultry related enterprises such as farms, breeding companies, and plants specialized in poultry rendering.

Production Process

The primary role of various breeding companies is to breed and maintain different breeds of birds which are either graded up or cross-bred to have hybrid chicks and poults. Most of the breeding chicks are usually selected based on various traits such as efficiency in poultry feeds conversion and ability to accumulate large amounts of meat in their breast. Most breeder companies rear breeder offspring to sexual maturity. When these breeder offspring are mature enough, they are used in production of fertile eggs. These eggs are then sold to various poultry integrators. Once the eggs are received by the integrators, they are then place in hatcheries for hatching. Fertilized turkeys and chicken eggs take 28 and 21 days, respectively, to hatch into new offspring.

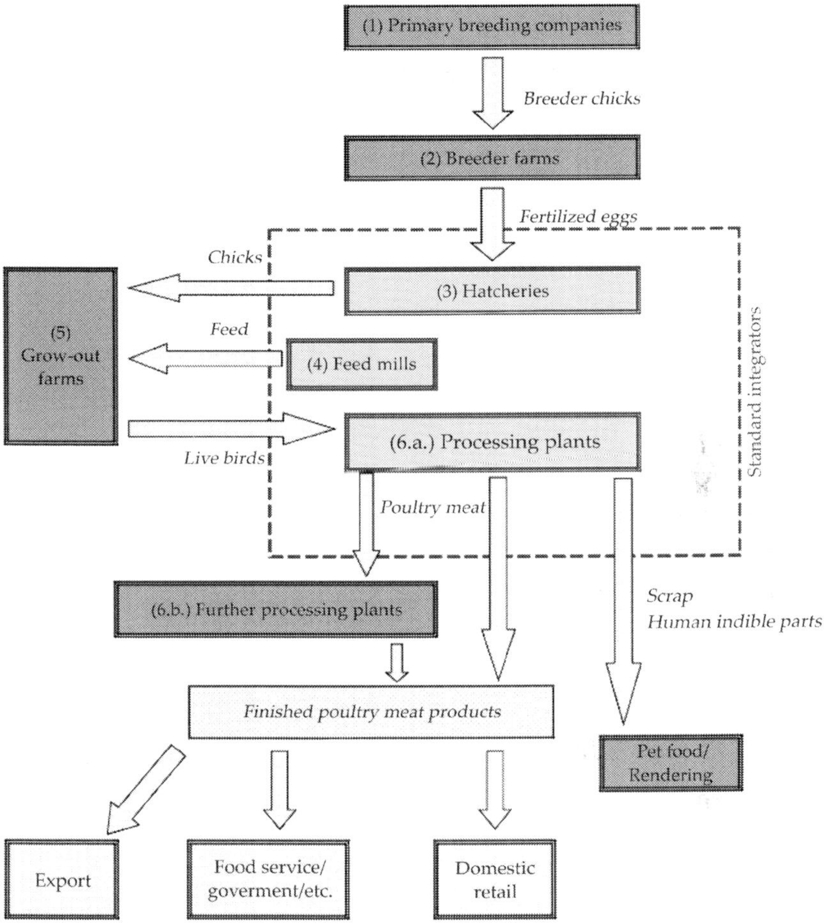

Figure 7.2: Poultry production process.

Source: *https://www.ers.usda.gov/data-products/chart-gallery/gallery/?page =39&topicId=0*

To prevent the embryo from sticking on the shells during incubation period, eggs are mechanically rotated periodically in the hatcheries. Vaccination is done to the embryo before and after hatching against various poultry diseases. Sorting of the eggs in hatcheries is usually done for the purpose of grow-out barns. For the production of broiler meat, both sex chicks are used and separated later by sex if specialized production such as male roaster or female Cornish Hens is required.

Feeds arc crucial elements in production process of poultry birds. This is because feeds help in improving the speed at which meat type poultry birds reach live-market weight. Apart from breeding, most poultry integrators own poultry feed mills in order to supply and control the type of feed given to their birds. Most poultry feeds are developed and formulated by various poultry nutritionists. Proper ingredients are used in formulating essential compositions of the feeds. There are three kinds of feeds for meat type birds-starter, grower, and finisher. Each kind has specific nutritious composition. Soybeans and corn are the main ingredients used in development of broiler and turkey feeds.

Most broiler and turkey are usually reared from chicks to the required live-market weight at grow-out farms. These farms are own and run by self-reliant farmers who are usually subcontracted by various poultry integrators. About 99% of all broilers for meat productions are reared by contracted farmers, while 80% of turkeys are reared under contractual means. The reason why poultry integrators usually prefer to subcontract poultry production operations to various farmers in due to high initial cost incurred in construction of the grow-out houses used to rear the birds. In other words, poultry integrators usually pay the farmers to raise their young birds until they reach maturity stage. By the year 2006, there were about 70,000 grow-out houses in the US for rearing broilers.

Despite subcontracting farmers to raise the birds for meat production, most services such as veterinary services and provision of poultry feeds are done by the integrators. On the other hand, contracted farmers are responsible for the provision of labour, electricity, litter, water and other managerial services. On reaching the require market live-weight, the birds are collected and then transported for processing in the integrators plants. The feather, feet, and head of the birds are removed at the integrators processing plants. Based on various federal or state regulations, inspection on the carcasses is done. Some treatments such as chlorine rinsing are done on raw poultry meat in order to reduce pathogens. After passing the inspection stage, the carcases are sorted for marketing either as whole fresh birds or in pieces. Packaging for the whole birds is done while the remaining parts are passed through other processing stages for cutting into pieces. Deboning of the carcases is also done mechanically. Some of the cut up pieces for sale include leg quarters, legs, wings, breast meat, and thigh meat.

Some processing poultry companies also undertake other activities such as branding, cooking, marinating, and nugget or sausage preparation. Additionally, poultry meats may also be incorporated into other products such as pot pies and microwaveable meals. Cooked turkey and ground turkey breast meats are also commonly used for preparing deli meat.

Market Channels

Poultry meat distribution is done through four key channels. Three of these channels are used in distribution of finished poultry products. These channels include export, such as government and food services, domestic, or local retail. The fourth channel involves rendering or pet food. Export distribution of the poultry products is directly managed by poultry integrators or indirectly handled and managed by the trading companies, distributors or brokers. In the US, local poultry distribution or sales are handled and managed directly by the poultry integrators or by the processors. Sometimes the sales are also handled and managed indirectly by the distributors, who purchase poultry products mainly from various integrators or further processors. The biggest poultry marketing channel in the year 2007 was local or domestic retailing mainly through grocery stores as illustrated in the next subtopic on production and consumption trends. However, exportation of broiler meat stabilized and overtook domestic retail in 2008 after rising to 19% from the previous 15% in 2006.

POULTRY MEAT PRODUCTION AND CONSUMPTION TRENDS

National Poultry Meat Production

The US reported around 9.2 billion live poults and broiler chicks newly hatched in the year 2012 for the production of commercial poultry meat as shown in the table below. This number actually indicates a decrease from the number recorded in the year 2006. It can be seen from the table that this number is actually the lowest from the year 2006 to 2012. The peak production of new poult and broiler chicks was recorded in 2007, but in 2008 there was a slight decline due to reduced poultry meat consumption that was attributed by economic recession in that year.

Table 7.3: US poultry production of live broiler and turkeys

Items	2006	2007	2008	2009	2010	2011	2012
Clicken: broiler chicks hatched	9,414	9,590	9,468	9,147	9,276	9,055	8,931
Turkey: poults	297	308	296	274	275	285	288
Total	9,711	9,898	9,764	9,421	9,551	9,340	9,219

Source: *https://www.ers.usda.gov/data-products/chart-gallery/gallery/?page=39&topicId=0*

Annual production of poultry meat (both turkeys and chicken broiler) was comparatively stable from 2006 to 2012. The highest production was recorded in 2008 (58.4 billion pounds), while the lowest poultry meat production was recorded in 2009 (54.9 billion pounds). There were fluctuations in poultry meat production every year with a slight growth of about 3% in 2008 and slight decline of 6% in 2009. It is evident enough that turkey and broiler meat production increased from the year 2006 to 2008, and this was mainly influenced by the high prices and high consumer demand for the poultry meat in 2007 and 2008. Successively, in 2009 there was a decline in poultry meat production attributed by the individual integrators in efforts to try and lower supply with an aim to boost the deteriorating market prices caused by economic recession that occurred in 2008–2009.

Table 7.4: US poultry broiler and turkey product by states (2006–2012)

Items	2006	2007	2008	2009	2010	2011	2012
Broilers							
Georgia	7,066	7,414	7,469	6,874	6,883	7,426	7,626
Arkansas	6 283	6,232	6,380	5,760	5,938	5,856	5,765
Alabama	5,688	5,625	5,846	5,513	5,787	5,820	5,620
North Carolina	5,099	5,390	5,493	5,317	5,419	5,587	5,678
Mississippi	4,662	4,614	4,876	4,602	4363	4,626	4,508
Texas	3,330	3,266	3,461	3,611	3,645	3,657	3,495
Kentucky	1,590	1,668	1,653	1,668	1,673	1,705	1,733

Maryland	1,607	1,592	1,612	1,401	1,433	1,649	1,611
South Carolina	1,408	1,480	1,516	1,522	1,557	1,542	1,537
Delaware	1,614	1,598	1,579	1,597	1,625	1,525	1,505
All other	10,482	10,452	10,556	9,877	10,427	10,684	10,452
U S broiler total	**48,829**	**49,331**	**50,441**	**47,752**	**49,153**	**50,077**	**49,530**
Turkeys							
North Carolina	1,125	1,147	1,208	1,090	969	1,030	1,177
Minnesota	1211	1,218	1,306	1,161	1,175	1,163	1,164
Indiana	462	481	519	543	574	578	606
Arkansas	585	572	611	568	562	607	577
Missouri	634	646	651	611	583	565	576
South Carolina	386	382	478	433	428	481	498
Virginia	555	483	484	449	449	460	449
California	428	450	435	390	409	422	431
Ohio	191	227	230	203	193	201	219
South Dakota	158	173	189	187	191	186	194
All other	1,729	1,787	1,811	1514	1,588	1,581	1,656
U S turkey total	**7,464**	**7 566**	**7,922**	**7,149**	**7 111**	**7,274**	**7,547**
U S broiler *and* turkey total	**56,293**	**56,897**	**58,363**	**54,901**	**56,264**	**57,351**	**57,077**

Source: *https://www.ers.usda.gov/data-products/chart-gallery/gallery/?page=39&topicId=0*

State Poultry Meat Production

The largest poultry meat production mainly come from the mid-Atlantic and south-eastern states as illustrated in the table above. States recording the highest production also tend to produce the highest live-bird. This is because most grow-out farms are mainly clustered around hatcheries of the poultry integrators.

There are specific states that account for the highest records on broiler meat production. From 2006 to 2012, the best five broiler meat producers were Georgia, which was followed by other small states like Arkansas, Alabama, North Carolina, and Mississippi. On average, they all accounted for about 59% of the total broiler meat produced in the entire US region. Over this period, these states recorded differently in terms of overall growth rate, with an increment of 8% in Georgia and 11% in North Carolina, and a decline of 8% in Arkansas and 3% in Mississippi. The peaks of broiler meat production by these five states were recorded in 2008, but there was a decline in production in 2009 due to the economic recession that occurred in 2008–2009.

Similarly, the highest turkey meat production in the US also comes from certain specific states. During the entire period from 2006 to 2012, the best two states that recorded the highest turkey meat production were Minnesota and North Carolina. These two countries actually accounted for more than a third of the total turkey meat produced in the US, while top five best countries accounted for around 54% of the total turkey meat production. There was a 4% decline in turkey meat production for Minnesota between 2006 and 2012. On the other hand, there was an increase of 5% in North Carolina over the same period of time.

The total turkey meat production grew through 2008 for both the two states (North Carolina and Minnesota), but then went down in 2009 because of the 2008–2009 economic recession. Similarly to other Atlantic region states, turkey meat production of North Carolina went further down in 2010. This was mainly attributed to integrators cutting down production, which included the two largest turkey producing companies (Cargill VAM and Butterball LLC), both operating entirely in the mid-Atlantic region. This was actually an initiative of integrators with an idea of restricting turkey meat supplies in order to enhance prices.

FACTORS AFFECTING POULTRY MEAT PRODUCTION

Some of the main inputs in production of poultry meat include breeding birds and eggs, poultry feeds, drinking water, housing, medicine, and labour among other inputs. Poultry meat processing plants and grow-out houses are assets that are not convertible into other uses. Among these inputs, feeds is the most unstable poultry production input based on the costs incurred in formulating and producing. High cost for producing poultry feeds is caused by exorbitant prices of the ingredients used such as soybeans and corn.

Production Costs

Poultry feeds, which is formulated primarily from soybeans and corns account for about 60–80% of the total production costs for poultry meat. Therefore, the costs of poultry feeds used for both turkey and broiler chicken rearing are directly influenced by the prices of these ingredients (soybean and corns). Availability and cost of these ingredients is also affected by various factors such as demand levels, environmental weather conditions, prices of fuel and energy, and other governmental policies. The figure below illustrates the prices of corns and soybeans from 2005 to 2012.

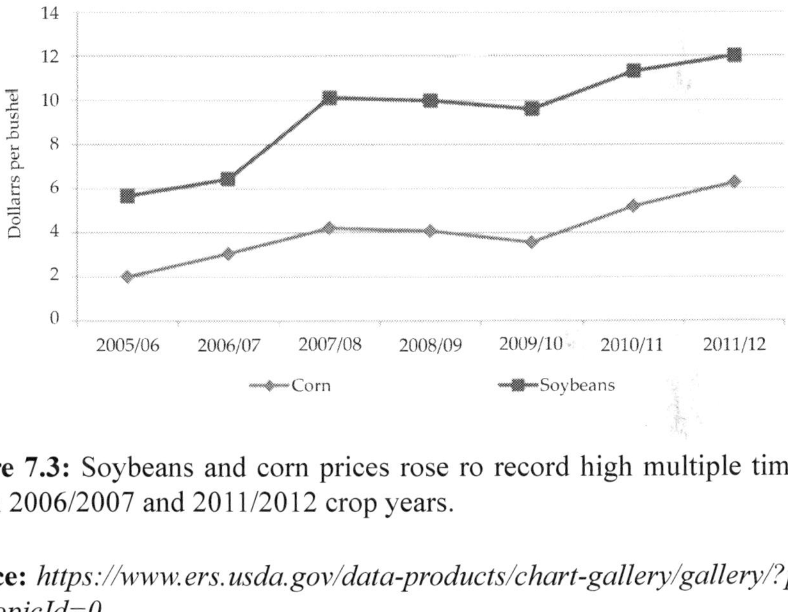

Figure 7.3: Soybeans and corn prices rose ro record high multiple times between 2006/2007 and 2011/2012 crop years.

Source: *https://www.ers.usda.gov/data-products/chart-gallery/gallery/?page=39&topicId=0*

The prices of corn and soybeans were high between the crop years 2005/2006 and 2007/2008. Because of the global economic recession, there was a decline in the prices of corn and soybeans between the crop year 2008/2009 and 2009/2010, but then rose again to new highs in the crop year 2011/2012. Most poultry integrators face huge loss in poultry meat production attributed by low domestic poultry meat consumption and high prices of the feeds. Most poultry integrators may be forced to respond to these effects by lowering poultry production, closing some of the production facilities, and consolidating various facilities that are inefficient with an aim of enhancing poultry meat production. According to poultry meat production industry representative viewpoints, if the prices of feeds are raised, it may cause the broiler and turkey integrators to sell their products at prices that do not cover their production costs.

Government Policies and Programs

Most production activities of poultry meat in the US are governed by federal laws and regulation in order to maintain food safety. One of the key federal laws is referred to as Poultry Product Inspection Act (PPIA). This law gives the US Department of Agriculture (USDA) the mandate to carry out inspection and monitoring various poultry products that are produced and sold locally and internationally. This law requires all inspected poultry producing facilities and processing plants to conduct both post- and ante-mortem to the birds prior to processing and maintain high level of sanitation during processing and packaging of various poultry meat products. The law also requires all poultry meat slaughter houses and processing plants to develop plans that will help in analysing, controlling, and reducing hazards related to food safety. In other words, law allow the processing plants to come up with a strategic plan that help in identifying any physical or biological hazards in the plants, developing critical measures for controlling and preventing occurrence of the hazards, monitoring all points tailored towards controlling those hazards, correcting all deviations and lastly maintaining critical plan records of all operations that happen in any poultry processing plant.

Research and Development

Research and Development in poultry meat production can actually be categorized into three key sections. First classification involves poultry genetics, poultry health, and poultry nutrition. The second classification involves technology in poultry production and processing. The third classification involves development of poultry products.

Research on poultry genetics is meant to help enhancing the quality of the poultry birds reared. This is an important area that helps most poultry breeders in production of first growing breeds, meaty breed, and healthy birds. Studies have shown that poultry genetic selection has always being responsible for about 80–90% of all first growing broiler chickens for the last few decades. The remaining 10–20% represent the improvement of poultry feeds. Continuous researches have also been done to help in improving the efficiency and the quality of various poultry feeds. These researches include researches on the use of different varieties of feed ingredients, on enzymes, the size of feed particles, and formulation. For instance, researches have been done on how poultry breeders can formulate poultry feed enriched with omega-3 fatty acids.

There has also be continuous researches on the various measures for controlling and preventing poultry disease occurrence, especially those that mainly attack live poultry birds such as clostridia dermatitis, spinal abscesses, avian influenza (AI), and other foot pad infections.

There have also been technologically tailored researches done with an aim of improving efficiency of poultry production process in all stages involved. This include ways on how to actually enhance cost effectiveness and other conditions of grow-out poultry houses, such as enhancing ventilation, cooling and heating systems in broiler houses. There have also been several efforts to improve efficiency in processing machineries through automating various sections of poultry processing that are done manually. For instance, there has been an effort on developing automated deboning machines by the Georgia Technology Research Institute.

Continuous researches have also been done on how to enhance waste and water management in poultry industry. These include ways on how water can be used more efficiently in poultry production and how wastewater can be recycled. On the development of poultry product, more focuses have been emphasized on various issues concerning the safety and health of products sold to the customers. These include ways of identifying spoilage on poultry products and suitable methods of avoiding those spoilages.

FACTORS AFFECTING POULTRY MEAT CONSUMPTION

One of the main factors affecting purchasing decision made by poultry meat consumers is the prices of the products. Based on studies, it has been revealed that an increase in price of one poultry product may actually lead to an increase in consumption of another poultry product substitute. For instance, many customers of poultry meat may be forced to shift to beef meat as a result of price increment on poultry meat. Alternatively, an increase in the price of poultry meat may force low income earners to shift their consumption to other poultry products such as eggs. Nevertheless, poultry meat is one of the lowest priced meats in most parts of the US and continues enjoying a huge demand, both locally and internationally. Based on the table below, it is important to note that in 2012, customer paid more than 45% for a pound of pork meat and more than 60% for a pound of beef than a pound of broiler chicken meat.

Table 7.5: Average retail prices for chicken, beef, pork, and egg (2006–2012)

Product	2006	2007	2008	2009	2010	2011	2012
	Dollars per pound (retail night equivalent) or per dozen eggs						
Chicken	1,57	1,65	1,75	1,75	1,75	1,77	1,89
Beef	3,97	4,16	4,33	4,33	4,40	4,83	5,02
Pork	2,81	2,87	2,87	2,94	3,11	3,43	3,47
Eggs (grade A, large)	1,31	1,68	1,68	1,99	1,66	1,77	1,84

Source: *https://www.ers.usda.gov/data-products/chart-gallery/gallery/?page=39&topicId=0*

Other factors affecting the demand and consumption trends in poultry meats include the lifestyles of the consumer, science, and other cultural traditions. For instance, improved awareness about health related issues of consuming white poultry meat (white meat is considered healthier and good than the red meat) has actually enhanced poultry production and consumption in many ways. Many people also want products that consume less time in preparing. For this reason therefore, many poultry producing and processing companies have actually developed various poultry product in order to meet this criteria. These include pan-ready poultry meals, pre-marinated poultry products, pre-packaged poultry meals, and other microwavable poultry products. These poultry products have really enhanced the demand and consumption rate of poultry meats. Seasonal demands have also affected consumption of turkey meats especially during holiday seasons such as end-of year and Easter holidays.

SECTION-IV
POULTRY GENETICS

CHAPTER 8

INTRODUCTION TO POULTRY GENETICS

FUNDAMENTAL ELEMENTS IN POULTRY GENETICS

Poultry breeders must have basic knowledge about poultry genetics, particularly understanding fundamental genetics elements and why different poultry birds have certain characteristics. There are terms commonly used in poultry genetics and breeding. It is important to first understand these terms before engaging into a poultry breeding business.

Chromosomes

There are thread-like structures or bodies observable in the cell nucleus of any animal. These cell nucleuses can be observed through special staining methods especially during mitosis metaphase stage. Chromosomes are made of deoxyribonucleic acid (DNA), which are usually characterised by double-stranded coil of helical molecules. Chromosomes are seen in pairs. The number of chromosomes in each pair is referred to as diploid (2n) number. Different poultry birds have different diploid numbers.

Table8.1: Illustrating various birds, their species, and diploid numbers that they possess

Common name of the birds	Species of the birds	Diploid Number (2n)
Chicken	*Gallus domesticus*	78
Duck Mallard Muscovy	 *Anas platyrhynchos* *Cairina moschata*	 80 80
Turkey	*Meleagris gallopavo*	80
Goose	*Cygnopsis cygnoides*	80
Japanese quail	*Coturnix japonica*	78
Pigeon	*Columba livia*	80
Guinea fowl	*Numida meleagris*	78
Ostrich	*Struthio camelus*	80

Source: *http://woolshed1.blogspot.co.ke/2009/05/introduction-to-practical-animal_17.html*

The main function of the chromosomes in poultry and other animal is to transmit hereditary materials from generation to another. Chromosomes can actually be classified into two main categories. These are the autosomes chromosomes and the sex chromosomes. Autosomes describe all other chromosomes except for the sex chromosomes. Sex chromosomes on the other hand are mainly for transmission of various genetic materials that are used in determining the sex of poultry offspring. For instance, human sex chromosomes are denoted by X and Y. Individuals with sex chromosomes denoted by XX are referred to as female individuals, while those with sex chromosomes denoted XY are referred to as male individuals. In poultry, sex chromosomes are denoted by Z and W. A chicken with sex chromosomes denoted by ZW are referred to as female chickens, while those with sex chromosomes denoted by ZZ are referred to as the male chickens. The difference between poultry chromosomes and mammal chromosomes is illustrated in Fig. 8.1.

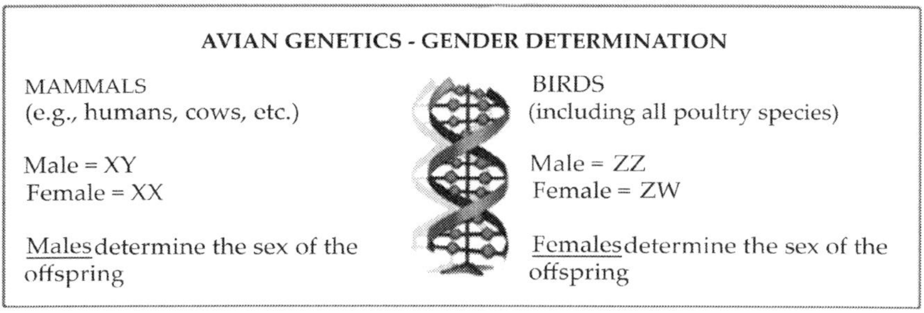

Figure 8.1: Sex chromosomes of animals and birds.

Source: *http://articles.extension.org/pages/65471/sex-linked-traits-in-poultry*

Genes

These are DNA segments used to carry chromosomes genetic materials. Genes are characterised by spiral-like structures resembling long ladders that appear as if they are twisted. There are hundreds of genes in a pair of chromosome. They are usually positioned in special location called locus with chromosomes. Because most genes are found in pairs within the cells of the body, a pair of genes can have certain effects and another pair of the same genes may have different effects on the same trait. Genes have different roles in the body of poultry. These functions include storing and transmitting information related to genetic materials, copying or replicating themselves with great precision and consistency, and undergoing mutation through replicating or copying themselves.

Because all chromosomes occur in pairs, so does the genes. Each poultry parent usually contributes one type of gene in every pair. The phenotype representing specific characteristic in a chicken is determined by the makeup of each pair of the genes for that specific trait. Homozygous is a term used to describe genetic state of genes having similar characteristics or traits. On the other hand heterozygous is a term used to describe genetic state of genes that have different characteristics. Those genes with both heterozygous and homozygous traits are commonly described as dominant factor.

There are three different conditions that may occur when dealing with characteristics where recessive genes and dominant genes are involved. The first condition is the homozygous dominancy condition which happens when both genes used in describing traits are dominant genes. The other condition is the homozygous recessive which happens when these genes used are both in recessive state. The third condition is the heterozygous dominant which happens when one of the genes used is recessive gene while the other gene is a dominant gene. In this case the two forms in each pair of genes are referred to as alleles.

The dominant trait genes in the heterozygous condition are usually expressed on the recessive genes. However, sometimes genes are actually capable of expressing some degrees of heterozygous condition in specific gene pairs. This occurrence is usually described as codominance, in which each pair of the genes can actually contribute equal expression, or one pair of the genes can dominate the contribution compared to other pairs.

Not all traits in poultry are controlled by specific gene pairs. Sometimes a particular trait of poultry can control by different types of gene pairs. If this happens, these traits are said to be quantitative. For instance, the brown colour on the shell of chicken eggs is usually controlled by as many as 13 genes. The range or different intensity of the brown colour seen in eggs is determined by the breed of the chickens.

OBSERVABLE GENETIC TRAITS IN POULTRY

Understanding why different poultry breeds have certain traits is very significant especially when breeding. Some of the observable traits in chickens and other poultry breeds include the colour of the feathers, the type of comb and wattles, skin colour, and the colour of foot and shank among other traits.

Genetics of Combs and Wattles

Different chickens possess different varieties of combs and wattles as shown in the figure below. Historically, genetic traits of chicken combs are significant trait when breeding. According to Gregory Johann Mendel, who is considered as the father of various genetic studies revealed that different physical traits of chickens and other animal are controlled by genes composition in their body. Based on this genetic idea, another scientist referred to as William Bateson applied genetics to illustrate why and how different birds of the same breed actually develop or have different traits on their combs and wattles.

Figure 8.2: Different combs types in poultry due to genetic variations.

Source: *http://articles.extension.org/pages/65363/poultry-genetics-for-small-and-backyard-flocks:-an-introduction*

The type of comb in poultry especially the chickens is usually predetermined by two genes belonging to two different chromosomes. One type of these genes is referred to as the rose comb gene (which is denoted by a letter *R*). The other type is referred to as the pea comb gene, which is denoted by the letter *P*. Uppercase letter is usually used to denote the presence of the comb type gene, while lowercase letter is usually used to denote absence of the comb type gene. An expression of both the pea type comb gene and rose type comb gene may occur, and is usually described as heterozygous state when it occurs. This means that the type of chicken comb can sufficiently be determined by either one copy of pea comb gene or one copy of rose comb gene. Therefore, both rose comb gene and pea comb gene can be described as a dominant gene in presence of at least a copy of one rose comb type gene and an absence of pea comb type gene. In this scenario the resulting expression is a rose comb. This means that chickens having rose combs have at least one of the two possible combinations of the genes (RRpp or Rrpp).

Another ways in which dominant genes can actually occur is when at least one of the pea comb type gene exists and there is absence of rose comb type gene. The resulting comb type is referred to as pea comb. For the chickens with pea comb, they may possess either of these two gene combination (rrPP or rrPp). Another case of dominant gene is referred to as walnut comb. This comb is usually developed when there is at least one copy of both pea comb genes and rose comb genes. Those chickens having walnut comb may possess at least one of the four possible combinations of the genes (RRPp, RrPp, RRPP, or RrPP). Lastly, the occurrence of dominant genes can happen when both rose comb type genes and pea comb type genes are absent. This usually results into a single comb. Chickens with this type of comb usually have only one gene combination possibility (rrpp).

	Rp	Rp
rP	RrPp	RrPp
rP	RrPp	RrPp

Figure 8.3: Gene combinations resulting from a pea comb and rose comb cross chickens breeding.

Source: *http://articles.extension.org/pages/65363/poultry-genetics-for-small-and-backyard-flocks:-an-introduction*

To clearly understand how comb types are influenced by the genetic combination in chickens, it is advisable to consider different result obtained after various chicken breeds are cross-bred. For instance, consider the possibility of cross breeding a chicken with a true pea comb type (a chicken having rrPP genes combination) with another chicken breed having a true rose comb type (a chicken having a RRpp genes combination). In this case, each chicken parent donates each of the two kinds of genes that predetermine the comb type for the offspring. For the development of a true rose comb type in offspring, the parents are supposed to donate only the Rp gene combination. Likewise, for the development of a true pea comb type in offspring, the parents used in breeding should donate rP gene combination. Subsequently, all young chicks obtained after cross breeding these parents would definitely possess heterozygous state for the two gene combinations (that is, RrPp) and all should develop walnut comb types. However, if offspring with walnut comb types are then cross bred, the resulting offspring will not have walnut combs. This means that if a bird with walnut comb is bred with another bird that has walnut comb type, the resulting offspring will develop other type of comb rather than the walnut comb type.

Based on the results in the above figure, if two offspring obtained are cross bred, the results would increase the number of possibilities for genes combinations. In this case, every parent would donate at least one of the four likely gene combinations. This would actually lead to development of 16 different offspring with different genes combinations. Based on the phenotype that is linked to every combination of the gene, one can actually be able to obtain particular odds of certain comb type that each offspring develops. In the figure below, it is clearly indicated that there would be possibility of having nine walnut comb types out of the 16 possible offspring obtained, three rose comb type possibilities, three pea comb type possibilities, and only one single comb type possibility.

	RP	Rp	rP	rp
RP	RRPP Walnut	RRPp Walnut	RrPP Walnut	RrPp Walnut
Rp	RRPp Walnut	RRpp Rose comb	RrPp Walnut	Rrpp Rose comb
rP	RrPP Walnut	RrPp Walnut	rrPP Pea comb	rrPp Pea comb
rp	RrPp Walnut	Rrpp Rose comb	rrPp Pea comb	Rrpp Single comb

Figure 8.4: Gene combination obtained after crossbreeding offspring in figure 8.3 above.

Source: *http://articles.extension.org/pages/65363/poultry-genetics-for-small-and-backyard-flocks:-an-introduction*

Genetics of Feather Colour

Before exploring genetic formation of feather colour in poultry, it is highly advisable to first get the idea on how different colours are achieved in poultry. Most domesticated birds have two main types of feather colours, which are primary feather colour and secondary feather colours. Secondary colours are those colour patterns that are usually seen on a single feather. On the other hand, the primary feather colours are those patterns of the feather usually seen on the whole body of the bird. For instance a chicken having Silver Columbian feather pattern is usually an ordinary white chicken possessing some black coloured feathers on the neck, tails areas, and the wings. Because the colour of the pattern is not noticeable on specific feathers, it is usually referred to as primary pattern.

To obtain chicken breeds that possess a certain colour pattern, it is important for the breeders to start with the contextual or background colour. This background colour scheme is influenced by a gene known as E-locus. This background colour pattern is usually modified by secondary and other colours. Essentially, the pattern and the colour of chickens' feathers are determined by combination of different genes. Bearing in mind the possibilities of having a white-coloured and black-coloured chicken, there are usually three distinct colours of feathers, which are red (gold) white and black. Theoretically, the black colour and white colours are not real colours. White colour on the feathers of chickens usually result from a combination of all different colours, while the black colour on the feathers is caused by lack of light reflection in the visible range.

Chickens obtain their feather colours through masking and enhancing the red and black colours. For instance, the colour on the feathers of Rhode Island Reds is usually influenced by the dominant mahogany gold gene (red-enhancing gene). The blue colour on the feathers of chickens is usually influenced by the blue genes, which are responsible for diluting the black colour. If two chickens with each copy of blue genes are crossbred, the resulting offspring will have a splash effects on the feathers. White-coloured chicken are obtained in several ways mainly through inhabitation of the red and black pigmentation on the feathers through genes combination. The results include Cuckoo barring, dominant white, Columbian, silver, and recessive white.

Sometimes the colour of the feather can actually be as a result of feather structure rather than its pigmentation. For instance, the beetle green and purple sheens observable on various poultry birds are usually as a result of light reflection on the structure of the feathers rather than pigmentation on the feathers.

Genetics of Foot/Shank Colour

The colour seen on chickens feet or shanks is as a result of colour combinations of the deeper skin and the upper skin of the chicken. There are two types of genes that control the colour on the feet or shanks of the chickens. One of these genes is referred to as sex-linked gene, which is usually found in the Z sex chromosomes of the chicken. The figure below clearly illustrates various colours of the feet/shanks of the chickens resulting from combination of major sex-linked genes of the chromosomes. While looking at these combinations, it is important to remember that every chicken possesses two different copies of each sex-linked gene. This diagram is therefore meant to provide a guideline because there are other types of genes that influence the colour of feet/ shanks in chickens. For instance, chickens have a sex-linked gene (B) which is usually found in the Z sex chromosomes. This sex-linked gene B acts as a strong melanin pigment inhibitor on the skin of the chickens. Sex-liked barring gene (B) is actually responsible for the light shanks on the Barred Plymouth Rock chickens. Most female Barred Plymouth Rock chickens usually have darker shanks compared to the males because they possess ZW sex chromosomes while the males only have ZZ sex chromosomes. This phenomenon is actually influenced by the effects on the dosage of the sex-linked barring genes in males.

Shank/Foot color	Genes
Near black with white soles	W^+ . Id. E
White shanks and feet	W^+ . Id. e^+
Black shanks. White soles	W^+ . Id^+. E
Blue shanks. White soles	W^+ . id^+. e^+
Near black with yellow sole	W . Id. E
Yellow shanks and feet	W . id. e^+
Black shanks with yellow soles	W . Id. E
Green shanks with yellow soles	W . id^+. e^+

Figure 8.5: The genetic of foot colour/shank.

Source: *http://articles.extension.org/pages/65363/poultry-genetics-for-small-and-backyard-flocks:-an-introduction*

Genetics of Dark Skin Colour

Most Silkie chicken breeds are usually known for their dark colour on their skin. This dark skin is caused by excessively high levels of melanin on the chicken's skin. Scientifically, this dark pigmentation on the skin of Silkie chickens is caused by fibromelanosis (which is a cell activator that influences pigmentation on the connective tissues). Dark skin phenotype inheritance on Silkie chickens is usually regulated by the fibromelanosis gene (denoted by letters Fm) and various inhibitors of dermal melanin (such as sex-linked ld dermal melanin-inhibiting mutation). Chickens that possess Fm genes in their sex chromosomes without the ld dermal melanin genes usually have dark colour on their skins and on various connective tissues. On the other hand, if the chickens have a combination of both the Fm genes and ld mutation usually do not have any pigmentation on their skins at all. The intensity of pigmentation (melanisation) resulting from Fm genes may also be influenced by other dermal melanin inhibitors. Studies have also shown that plumage colour inhibiting genes (such as E-locus alleles) can sometimes influence the colour of the chicken's skin. This results due to the expression influence of Fm gene by the other genes.

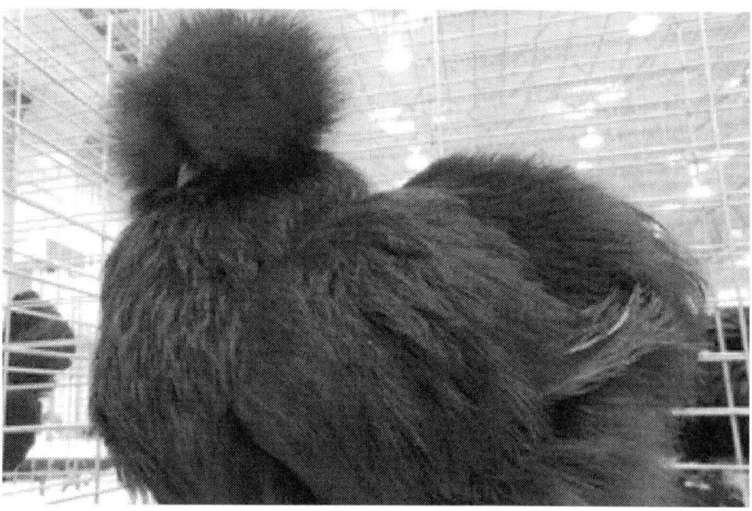

Figure 8.6: Genetics of dark skin colour in poultry.

Source: *http://articles.extension.org/pages/65363/poultry-genetics-for-small-and-backyard-flocks:-an-introduction*

SEX-LINKED TRAITS IN POULTRY GENETICS

Genetically, the gender principle of any animal is usually determined by two different sex chromosomes. In most of the mammals, these sex chromosomes are usually denoted by the letters X and Y, whereby males possess the XY chromosomes and the females possessing the XX chromosomes. Therefore, the gender of the offspring is genetically influenced by the XY chromosomes (that is the males). In poultry, conversely, the gender principle of the birds is usually determined by the sex chromosomes Z and W, whereby the females possess the ZW chromosomes while the males possess the ZZ chromosomes. It is therefore clearly shown that the gender of poultry offspring is usually determined or influenced by the ZW sex chromosomes (that is the females).

Apart from influencing the gender of the offspring, poultry sex chromosomes are also responsible of conveying other genes that influence other different chickens' traits, such as the colour of the plumage. These traits have actually been put into use in sex-linked cross-breeding of the poultry, therefore influencing the gender of different offspring in poultry. Studies have also shown that based on specific combination of sex-linked cross-breeding, males and females of differing poultry breeds can actually be mated to obtain offspring with traits of cross-bred birds. For instance, if a Delaware female chicken is cross-bred or mated to male New Hampshire, the resulting offspring (males) will actually mature to possess feather pattern resembling those of the female Delaware mother, and other offspring (females) will actually mature to possess feather pattern resembling those of their male New Hampshire father.

Consequently, if a male Delaware chicken is mated to a female New Hampshire female, all the resulting offspring will actually possess similar feather pattern resembling those of their Male Delaware father. When doing poultry breeding, it is essential to always remember that offspring obtained from a sex-linked cross-breeding cannot be used in the second sex-linked cross-breeding.

Sex-Linked Crossing Breeding In Layers

There are two main sex-liked crosses in egg production, which are the black sex-links and the red sex-links. For the black sex-links also commonly referred to as the Rock Reds are obtained through cross-breeding barred female chickens with non-barred male chickens. The male chicks usually have same feathers resembling those of their barred mother, while on the other side female chicks will develop solid coloured feathers resembling those of their non-barred father (usually black). One good example of a sex-link cross-breeding involves Barred Plymouth Rock female chicken mated to a New Hampshire or Rhode Island Red male chicken. On hatching, all the offspring will have black down; however, the male chicks can be sorted out from the flock through identifying white marks or dots on their heads. Most female chicken obtained through sex-linked cross-breeding usually produce eggs that have brown shells.

Red sex-links offspring can be produced using numerous crosses. For instance, a white Plymouth Rock female chicken (having silver factors) can be cross-bred with another New Hampshire male chicken in order to obtain a gold comet. White Plymouth Rock Hens usually have silver genes in their sex chromosomes that influence red colour on their feathers. Another example involves cross-breeding a Wyandotte (silver-laced) female chicken with a New Hampshire male chicken to obtain cinnamon queen. Most of the modern commercial breeders for egg production are using the knowledge of sex-linkage through strains of white-shelled eggs in differentiating pullets from the cockerels when newly hatched chicks are one-day old. Additionally, the same technique is being used in determining rapid-feathering and slow-feathering sex-linked breeds in production of meat type chickens.

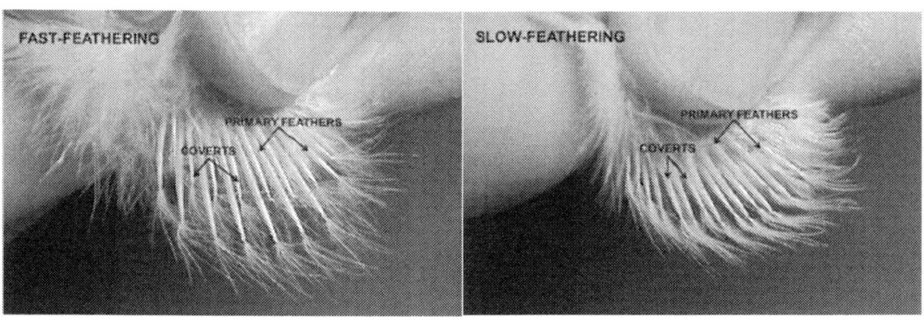

Figure 8.7: Rapid-feathering and slow-feathering sex-linked breeds in production of meat type chickens

Source: *http://articles.extension.org/pages/65471/sex-linked-traits-in-poultry*

Sex-Linked Crossing Breeding In Broilers

The rate at which the feathers of newly hatched chicks grow is usually related to sex-linked characteristic. This characteristic is used by breeders to identify and sort day-old meat type chickens based on their sexes. Slow feathering is a trait in broiler chickens that is influenced by a certain gene located in Z chromosomes. After chicks are hatched, the size or length difference of the feathers on covert and primary wings is usually observable within a period of three to five days. During this period, breeders can actually use this sex-link cross trait in sorting the chicks based on their gender. If a male with fast growing feathers is cross-bred with female with slow growing feathers, the female offspring will have fast growing feather like their father while the male offspring will have slow-growing feathers just like their mother.

GENETICS CAUSES OF "HEN-FEATHERING" IN MALE CHICKENS

Sometimes there are incidences when male chickens usually develop plumage with characteristics similar to those of female chickens. This phenomenon is referred to as Hen-feathering. A good example is when a male chicken develops feathers that are rounded rather than developing the usual male feathers that are normally pointed. Some of the common chickens that exhibit traits of hen feathering include the Campine and the Sebright chickens breeds as shown below.

Figure 8.8: Hen-feathering in male chickens.

Source: *http://forum.backyardpoultry.com/viewtopic.php?f=10&t=8007710*

Secondary sexual traits development in poultry as the birds mature is caused by production of hormones in the ovary of the females and testes of the males birds. The characteristics expressed by the birds are usually predetermined by the balance between production of oestrogen and androgen. The pattern and type of plumage of the female chickens is determined by the production of oestrogen hormones (feminize). This hormone is responsible for the formation and development of feather follicles that directly influence the roundedness of female feathers in the tail and hackles. This hormone is also responsible for pigmentation on the females' feather tracts. Chickens that express hen-feathering traits in males usually have a single gene mutation that cause over production of aromatase enzyme in various body tissues, including the birds' feather follicles. This enzyme usually helps in androgens conversion into oestrogens, resulting to excess amount of oestrogen in the feather follicles of Campine and Sebright chicken breeds. In other words, this means that the level of oestrogen hormones in the feather follicles becoming abundant to feminize feather growth on male chickens.

Androgens on the other hand are male hormones that enhance formation and maintenance of various characteristics in male chickens. One of the most commonly known androgen hormone is referred to as testosterone. These hormones are responsible for the inducement for the formation and growth of wattles and combs in male chickens. The hormones are also responsible for the crowing by the rosters. However, oestrogen and androgens hormones are not responsible for the formation and the growth of the spurs. Instead, growth and development of spurs in chicken is influenced by the genetic sex of the birds.

GENETICS SEXING OF DAY-OLD CHICKS

In most animals, it is actually simple to tell the gender of the new-borns. This is because most of the reproductive organs are usually positioned on the surface of the animals' body, which relatively make it much easy to see and locate them even in newly born young ones. However, this is extremely difficult in poultry because the reproductive organs of the males are usually located inside the body cavity. This means that sexing newly hatched chicks is not possible. However, with new advancements in researches, there are three methods that have been established that can actually be used in sexing newly hatched chicks at the hatcheries, which include the vent sexing method, feather sexing method, and sex-linked crosses.

Vent Sexing

This technique was actually initiated in Japan and later adopted by poultry bleeders in America in late 1930s. This is actually a skill that every poultry breeder adopts, although it usually takes time to develop. When practising this sexing technique in poultry, one should hold the chick upside down using one hand and then try to expel the faecal materials out of the vent area of the chick. When doing so, the bleeders should observe availability or absence of some rudimentary sex organs for the males. Although this technique may sound easier to perform, it actually requires somc skills, especially when identifying the sex organ. Therefore, one must be trained well and perform more practices in order to accurately perform chick sexing.

Feather Sexing

Using feather sexing technique, it is easier to sex certain chicken breeds after they are newly hatched. Some of the best chickens' breeds for feather sexing include the New Hampshire and the Rhode Island Red chickens. These breeds can be sexed using their physical traits such as the colour of their wings and feathers. Most male chicks usually have some white spots on their wings down. As the chicks age, the wings down is usually shed off and replaced with feather. However, chick sexing using this technique is usually not very accurate because of the various variations in spot size.

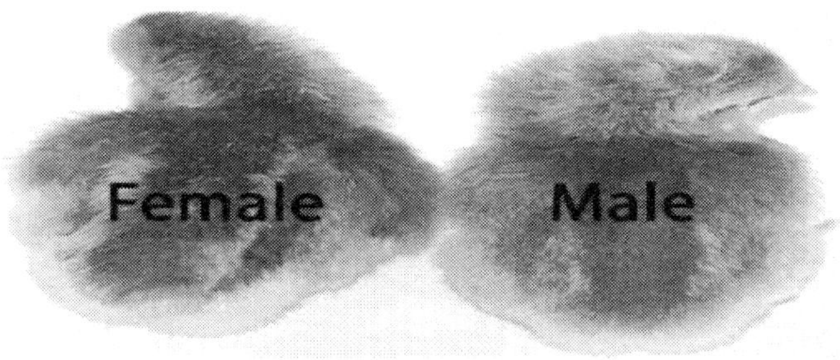

Figure 8.9: Feather sexing technique used in day-old-chicks.

Source: *http://countrysidenetwork.com/daily/poultry/chickens-101/how-to-tell-the-sex-of-baby-chicks/*

Feather sexing can also be practised in Barred Plymouth Rock chicken breeds. This is because they possess a barring gene (B) in the Z sex chromosomes as indicated in the figure below.

B = Barred b = non-barred

FEMALE

MALE		Z^B	W
	Z^B	$Z^B Z^B$	$Z^B W$
	Z^B	$Z^B Z^B$	$Z^B W$

$Z^B Z^B$ = Barred male with board white bars
$Z^B W$ = Barred female with more narrowhite bars

Figure 8.10: Genetic basis for sexing day-old Barred Plymouth Rock Chick.

Source: *http://articles.extension.org/pages/65437/sexing-day-old-chicks-on-small-and-backyard-flocks*

This barring gene (B) is responsible for the production of the white bar on the black feathers of the birds. Usually, this barring gene (B) is partly dominant over the other non-barring gene (b). Male adult birds possessing two barring genes (BB) usually form or develop white bars that are broader than the white bars developed on the female adult birds. This is because female adult birds only possess one of the barring genes (B_). Because of this difference, most females always appear lighter in colour than the males. The sex of chicks obtained through breeding pure Barred Plymouth Rock breeds can be identified using the shape and the size of the white (light-coloured) spot located on the head top. Male chicks usually have a large spots, while the white spot on female chicks is usually narrower and much small. Studies have shown that this sexing technique gives an accuracy of about 85% if it is used well.

Sex-Linked Crosses

Sex-linked crossbreeding is another technique that can be used in sexing chicks at hatching. This can be achieved using the characteristics of the feather colour. Colour traits on the feathers of chickens are determined by certain genes located on the sex chromosomes of the chickens. Therefore, in order to get understanding how sexing is done using this technique, it is important to gain knowledge on how genetics are involved. As discussed in other topic, half of the sex chromosomes in an offspring are usually contributed by the female mother bird while the other half is contributed by the male parent. The female mother bird usually contributes ZW sex chromosomes while the male parent bird contributes ZZ sex chromosomes. In this case, it is the female birds that determine the sex of the offspring in poultry.

Cross breeding various poultry breeds can actually result to formation of different characteristics of the offspring feathers. When a barred female parent is crossbred with another non-barred male parent, the female offspring are usually non-barred just like their male parent, while on the other hand the male offspring are usually barred just like their female parent as shown in the figure below.

B = Barred b = non-barred

FEMALE

MALE	Z^B	W
Z^b	Z^B Z^b	Z^b W
Z^b	Z^B Z^b	Z^b W

$Z^B Z^b$ = Barred male
$Z^b W$ = Non-barred female

Figure 8.11: Genetic basis for sex-linked of barred female to non-barred male.

Source: *http://articles.extension.org/pages/65437/sexing-day-old-chicks-on-small-and-backyard-flocks*

When they are newly hatched, all chicks (both males and females) usually have down that are dark in colour. The male chicks usually have a large white spot on their head tops. When using this sex-linked trait in sexing the chicks, it is advisable not to cross breed barred male parents with non-barred female parent because it won't work. Some of the commonly used chicken breeds include the non-barred male New Hampshire or Rhode Island Red.

Another genetic trait that has been used often in various commercial breeds is the silver (S) colour genes and the gold (s) colour genes. When silver females are mated to gold males, the offspring obtained are usually silver male offspring and gold female offspring as shown in the following figure. This genetic trait of gold and silver colour genes has often been used in sexing chicks both in broiler chickens and layer chickens.

S = silver s = gold

FEMALE

MALE		Z^S	W
	Z^S	$Z^S Z^S$	$Z^S W$
	Z^S	$Z^S Z^S$	$Z^S W$

$Z^S Z^S$ = Silver male
$Z^S W$ = Gold female

Figure 8.12: Genetic basis for sex-linked cross of silver/golden genes.

Source: *http://articles.extension.org/pages/65437/sexing-day-old-chicks-on-small-and-backyard-flocks*

Some breeds producing eggs with brown shells usually have gold or silver sexing traits. In meat-type chickens, one-day old female chicks usually have buff or gold gown while one-day-old males usually have white or yellow down. As the chicks grow, the females usually feather out the white and gold colour, but not the gold colour on the outer part of the feathers. Most often the quills and the undercoat of the male broiler chicks are white when processed.

Commercial egg laying producers are cross breeding the barred/non-barred genes and silver/gold colour genes with an aim of obtaining brown-shelled eggs. For instance, when a gold non-barred Rhode Island Red male chickens is crossbred with a silver Barred Plymouth Rock female chicken, the female offspring obtained are usually red/black in colour and non-barred, while the male offspring are usually white/black and barred. This is illustrated in the figure bellow.

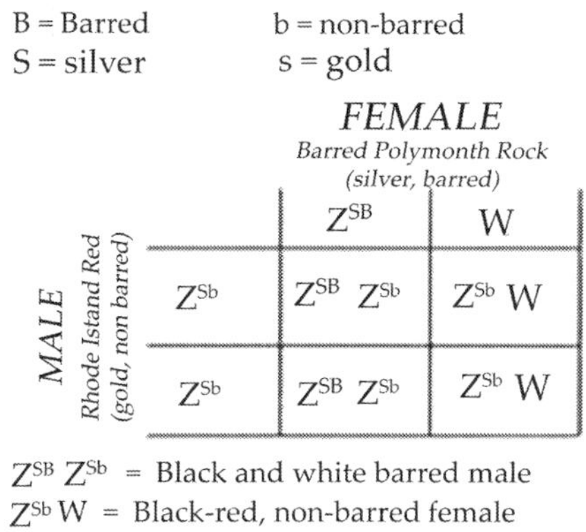

Figure 8.13: Genetic basis for sex-linked cross of silver/gold with barring gencs.

Source: *http://articles.extension.org/pages/65437/sexing-day-old-chicks-on-small-and-backyard-flocks*

Another sex-linked trait that is commonly used in sexing one-day-old chicks in commercial breeding companies is the fast-feathering gene. For instance, when slow-feathering female chicken is crossbred with a fast-feathering male chicken as shown in the figure below, the resulting offspring will have slow-feathering male chicks and fast-feathering female chicks. Therefore, the female chicks will have relatively longer feathers than the male chicks. The covert feathers of the females are however shorter than the primary feathers, while the male chicks have longer covert feathers than their primary feathers. To be able to accurately use the length of wing-feather in sexing the chicks, it is important to acquire some training from skilled poultry breeders. Nevertheless, the required training is usually minimal compared to the skill required in vent sexing.

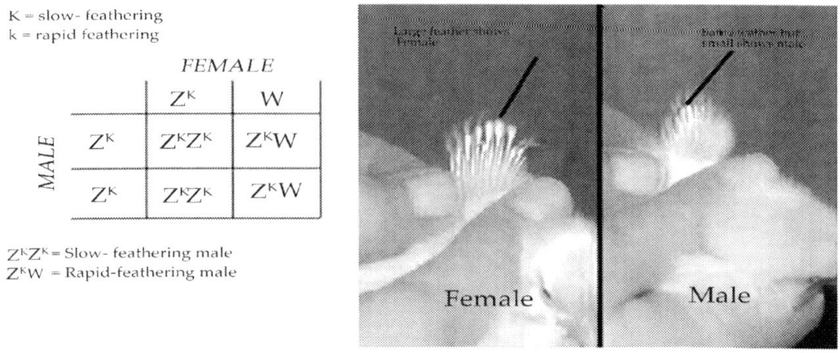

Figure 8.14: Genetic basis for sex-linked cross of rapid/slow feathering gene.

Source: *http://articles.extension.org/pages/65437/sexing-day-old-chicks-on-small-and-backyard-flocks*

In summary, as the chicks grow bigger, it becomes difficult to actually sex the chicks based on their physical characteristics. After they are newly hatched, both the male and female chicks usually have the same body weight, except for the differences mentioned above on their feather length differences. Neither the female chicks nor the male chicks exhibit any secondary sexual traits. Therefore, the best way to distinguish between female chicks from the male chicks is to actually watch them as they grow big. This is because as they grow bigger, the female chicks will exhibit different traits on their feathering and behaviours compared to the male chicks. These differences are described below

Table 8.2: Different traits on feathering and behaviours compared to the male and female chickens

Male chickens	Female chickens
• Have large wattles, combs, and bodies	• Have small wattles, combs, and bodies
• The comb will stand upright in case of a single-comb trait birds like the Leghorns	• Will have flopping comb over on one side in case of a single-comb trait birds like the Leghorns
• Exhibit large spurs	• Exhibits small spurs

• Males have more pointed, longer, and narrower hackle feathers (usually found on the neck)	• The hackle feathers are usually round and oval in shape
• Have saddle feathers on the tail	• Lack the saddle feathers on the tail
• The crest feathers for some crested chickens such as the Crevecoeur, Sultans, and Polish are usually pointed thus giving a punk-like appearance	• The crest feathers are mainly curved and have a topknot that is soft
• Males combs develop early and grow more fast than the female comb	• Female comb usually grow late and a slower speed than male combs
• They will tend to switch from chirping, which is a trait seen in all chicks, to trying to crow	

Source: *http://articles.extension.org/pages/66114/how-can-i-distinguish-male-and-female-chickens*

However, there are some exceptions when it comes to sexing day old chicks based on the physical traits in some chicken breeds. For instance, the Sebright and Campine male breeds usually have plumage that looks exactly like the female plumage. These breeds (males) are usually said to be "hen feathered." This genetic trait actually makes sexing so difficult for these chicken breeds. Another breed that is difficult to sex based on the physical genetic traits is the Silkie chickens. They do not have barbs (they look as if they have hair instead of feathers). Differentiating this breed based on sex-related traits in their physical appearance on the sickle and hackle feather is extremely difficult.

GENETIC SEX REVERSAL IN POULTRY

Studies have shown that sometimes female birds develop external traits of male poultry birds. This is a common phenomenon that occurs mostly in chicken and is usually known as sex reversal. A female chicken that anatomically and genetically experience some sex reversals usually exhibits its original feminine status, but most of its secondary physiognomies are similar to those of the male chickens. Typically, there are numerous differences that can be seen in the secondary sex traits of the male and female chickens. These differences include the following: male chickens usually have larger wattles, combs and general body conformation than the female chickens. For the single-comb chickens such as the Leghorn, the comb of the males is usually erect and turgid, while for the females, the comb usually appear flopping on one side of the head. The male chickens usually have a more developed and large spur compared to the female chickens. Although both the female and male chickens usually have the ability to crow, however this trait is prominently observable in male chickens. Based on the multi-colouration in chicken, most male chickens usually have more colour varieties for their plumage than the female chickens, with an exception for those chicken breeds where males are "hen feathered" such as the case of Campine and Sebright chicken breeds. The hackles of the male chickens are usually more pointed compared to those of female chickens. Both male and female chickens have the tail feathers, but sickles and saddle feathers are usually observable in male chickens.

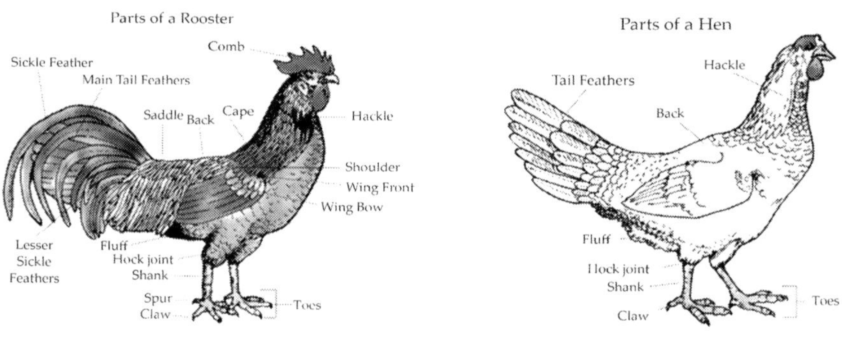

Figure 8.15: Genetic Sex Reversal in poultry

Source: *https://en.engormix.com/poultry-industry/articles/chicken-sex-reversal-t34107.htm*

Spontaneous sex reversal in is chicken occur in a weird way. Most female chickens usually have only one ovary that is functional (the ovary). The right chicken ovary and oviduct are usually observable during the embryonic phase of all poultry birds' development but they do not develop as the birds mature. Most spontaneous cases of sex reversal are caused by an infection condition that causes damage on the left ovary. In other words, sex reversal is as a result of regression of the left ovary resulting from a tumour, ovarian cyst, or an infection on the adrenal glands.

The residual tissues of the right ovary genetically develop without the presence of the functional left ovary. This development causes regeneration of a gonad on the right ovary commonly referred to as an ovotestis, which may have some tissue traits of testes, ovary, or both organs. Genetically, these gonad organs usually produce or secrete both oestrogens and androgens. Consequently, the female chickens usually form and develop various secondary sex traits resembling those of male chickens. Although the genotypic traits of the birds are for the female, the phenotypic traits exhibited by the birds resemble those of the male chickens (genotypic traits are those characteristics with actual codes for paired genes, whereas phenotypic traits refer to the secondary observable characteristic on the birds). Researches have revealed that some female chickens develop ovotestes with capability of semen production, thus making the birds to father offspring. However, most poultry birds that develop or experience sex reversal usually never sire offspring or lay eggs.

CHAPTER 9

GENETIC SELECTION AND MULTIPLICATION PROCESS FOR MEAT PRODUCTION

A BRIEF OVERVIEW ON BROILER SELECTION AND PRODUCTION

Production of broiler chickens is currently one of the biggest sectors in poultry industry. It is estimated that about 420 million parent broiler stocks are consumed annually in the whole world. According to USDA review, the US alone generate more than 44 billion US dollars annually from retail sales of the broiler meat. Although most commercial breeders for broiler meat production were initiated in late 1920s, adoption of genetically modified broiler stocks for commercial meat production did not start until in the early 1950s.

Efficiency and the quality improvements on breeding strains have always paid off. Since genetic selection strategies were adopted, the estimated cost for the production of one pound of live broiler chicken went down from US$2.4 in the year 1934 to about US$1.08 in the year 1960. According to a reported done by USDA on Poultry Yearbook in 2006, revealed that the cost for producing a pound of broiler live broiler chicken had actually dropped by about 50 cents by 2004.

Within broiler flocks themselves, it is evident to say that genetic selection of the breeding birds has really enhanced performance of the birds. This has been achieved through improvement on dietary, veterinary services, and control on various environmental conditions that influence the health and performance of the birds.

EVOLVEMENT OF GENETIC SELECTION CRITERIA IN BROILER INDUSTRY

In 1945, a breeding contests referred to as "Chicken of Tomorrow" was organised by the American Grocer A&P. Trials for contest qualification were then conducted in 1946 to 1947, while national finals held in the year 1948. During the national finals, all the broiler breeders that took part in the contest submitted a case of 30 dozen eggs to an Eastern Shore hatchery for hatching. The eggs were hatched and the young chicks given appropriate feed until they attained the required live-market weight. The chickens were then slaughtered. Different factors were put into consideration such as growth rate, efficiency on food conversion, and the quantity of the drumsticks and breast meat accumulated by the broiler chickens. Although the contests were held only on three different occasions, most broiler breeders such as Cobb, Vantress, Arbor Acres, Peterson, Pilch, and Hubbard became recognised market brands through these contents.

The first broiler breeders were highly dependent on mass selection (selecting individual broiler birds based on their characteristics purity). Because the main aim of most broiler producers was to select those first growing breeds, most broiler breeders used weight (choose heaviest females and males breeds) as their primary or main selection criterion. It was also revealed that the broiler weight was moderately heritable. In other words, 20 to 40% of all characteristics were genetically regulated. Although this selection criterion was somehow simple, it assisted many breeders in improving the performance of their birds across several generations.

Over several decades, selection criteria for broiler chickens have really been altered by the market forces. A more consolidated and integrated poultry industry understood that selection of breeding chicken based on their growth rates and weight gain alone could not sufficiently enhance production of meat in poultry industry. This fact was actually true with consideration on the exorbitantly rising poultry feed prices every year. Currently, broiler and other poultry feeds accounts for about 70–80% of the total costs incurred in production of broiler chickens from day one until their they acquire the right live-market weight. Because of this, selection for broiler breeding chickens based on the efficiency of feed conversion became a must criterion.

As broiler breeders became well conversant with various negative correlations that occur between reproductive and growth rate traits of the breeds, those aiming at cost effectiveness of broiler chicks production started assessing other selection criteria such as hatchability and improved egg production. Most breeders saw the essence in these selection criteria, thus causing many to stop the use of mass selection. Adoption of Selection Index theory was initiated, where breeders were able to select their breeding strains based on the information obtained from the relatives of the selected strains.

CONSOLIDATION OF GENETIC SELECTION IN THE '80S AND '90S

European countries and the US, which currently form the main markets for poultry meat, became mature in the production of chickens meat in the 1980s and 1990s. This led to more chicken meat retailers shifting from marketing disembowelled broiler carcasses to trap packed meat. This trend actually led to the production of more chicken meat components than the purely whole chicken production. As the market for chicken meat advanced, most retailers went on shifting from one aspect of broiler meat to another. Fresh tray-packed meat shifted to processed chicken meat products, with more emphases tailored towards enhancing production of breast meat in different markets.

During this period, most westernized poultry meat industries also became progressively focussed on enhancing food safety and the welfare of the breeds used in poultry meat production. This resulted to adoption of more complex selection criteria and objectives required of all breeders. There were different programs for selecting the breeding strain incorporated such as the Best Linear Unbiased Prediction (BLUP) broiler selection tools. This tool helped in weighing performance of breeding birds and their relative in order to assess how their characteristics related to one another. Through the use of BLUP, breeders were able to select those breeds that were better than others. Based on these variations, primary breeder industries consolidated with an aim to lower the high levels of research and development spending and other operative cost incurred within the poultry breeding business. By 2000s, there were only three poultry breeding group in operation, which include Aviagen (focusing on Arbor Acres, Ross, Peterson and Lohmann Indian River brands), Cobb-Vantress (mainly dealing with Avian, Hybro, Cobb, and Sasso brands), and the Groupe Grimaud (dealing with Grimaud Frere and Hubbard brands).

The history of broiler chicken production has influenced various criteria been used currently in the industry. With more priorities focussed on the efficiency of poultry feed conversion and growth rate of the breeds been used. The idea of producing broiler chicken breeds using economically viable number of chicks and eggs is considered an important element in broiler production. Both the quality and quantity of broiler meat production for the current breeders are also crucial elements that have contributed positively towards the continuous growth and success of poultry industry.

SUSTAINABILITY OF BROILER MEAT PRODUCTION

Currently, most broiler meat companies are focusing more on sustainability of broiler meat production than any other element. The fear more broiler breeders are currently having is whether the current breeding and genetic selection programs will actually continue to sustain production of purpose-fit meat type breeds in order to face increasingly scarce and costly poultry production resource. Numerous cases of leg related disorders problems, poor feeds consumption, poor immune responses, and high waste product output are currently not accepted in the production of broiler meat for the purposes of sustaining long-term breeding programs.

In broiler meat production, there are four key issues that influence the sustainability based on the selection and breeding programs adopted: these include

- Breeding programs that safeguard the health of human beings

- Breeding programs that ensure the welfare and health condition of the bred birds are safeguarded
- Breeding programs that ensure that customer financial success is sustained
- Breeding programs that enhance the long-term broiler industry sustainability

According to a reported given in United Kingdom in the year 2008, revealed that most commercially reared meat type chickens (broiler) usually emit the lowest volume of greenhouse gas in every one kilogram of produced meat compared to other livestock reared for meat production. This trait has actually been facilitated by the success achieved in broiler meat production especially on genetic selection of breeding strains in the past 20 years. Genetic selection has led to about 25% reduction in greenhouse emission by broiler chicken. With the on-going success and achievement in poultry industry, this reduction of greenhouse emission is predicated to go on for approximately the next few decades.

Most commercial poultry geneticists have also considered this as a conservative estimate bearing in mind the pace at which technological advancement in poultry industry are occurring. Based on genetic selection of the breeds for enhanced feed conversion efficiency have actually led to this reduction. There has also been a great reduction in the broiler meat wholesale price especially for the healthy and high quality animal protein to the end customers.

Improvement in the welfare of the poultry birds used in production of meat has also been influenced through various sustainability efforts. Although there have been some reports of several cases of physiological health conditions and physical disorders on genetically selected broilers, it is important to note that these problems are less prevalent than they were in the 1980s. With the use of advanced genetic technologies and other medically test tools to enhance selection strategies for broilers in the past 20 years, most of these physiological and physical incidences have been reduced to the lowest levels within effectively managed and controlled commercial broiler production sites.

Although it is difficult to measure the effectiveness on the reduction of these issues, several datasets results can be retrieved from different source such as Canadian Meat Inspection Service to help in illustrating how various physiological and physical ailments in current broiler chickens have been improved through genetic strength. Nevertheless, the rate at which these improvements are achieved cannot be accredited to the use of genetic selection alone, but also to the achievement acquired in the use of effective poultry biosecurity measures and other management practices, which directly influence the general production and performance of individual birds.

Globally, most broiler breeders are still pursuing various researches and poultry development initiatives designed to help in addressing sustainability and the welfare concerns of the birds. This enables various breeders to actually respond to various emerging poultry selection and breeding trends as fast as they can. However, because of the complex nature of selection and multiplication structure of broiler industry, there is usually a genetic drag of about 5–6 years from the time the first issue was addressed to the time the effects of the genetic selection can actually be noticed in the field. In other words improvements in broiler production are usually incremental, but they always work like "compound interest" for the dramatic changes to be felt over a period of time.

For instance, according to European Broiler Directive report, foot pad problem in broiler chicken is presently seen as one of the major broiler welfare issues. It has actually taken geneticists a few decades to come up with the best genetically viable selection criterion that will help in improving the quality of foot pad in order to alleviate various health related problems. However, because of genetic drag/delay broiler producers were forced to come up with other strategies that may help in managing and improving the condition of foot pad as the genetic selection and improvement progresses through the breeding generational pyramid.

GLOBAL AWARENESS AND FUTURE TRENDS OF GENETICS IN POULTRY MEAT INDUSTRY

In most developed countries like the US, Europe, Australia, New Zealand, and other countries, major poultry breeders are focusing more on improvement of poultry performance through leveraging various strategies tailored towards veterinary medicine, poultry nutrition, and controlling various environmental factors that impact on birds performance. This has actually been achieved through genetic selection of individual pedigree breeds for breeding under close to ideal environmental conditions. Certainly, countries like the US and New Zealand are on the forefront lane in production of broiler chickens with the best feed efficiency and highest growth rates. This has actually been facilitated by few or none of the endemic poultry infections in these countries, and also due to the continuous attention put by various broiler producers on information pertaining proper poultry management.

Different experts in poultry industry have for a long time discussed the likelihood of reaching to appoint where all biological barriers experienced in poultry production will be controllable to enhance poultry performance. However, because of high variation levels in poultry genome, it will take sometimes before these barriers are completely dealt with. Some countries like New Zealand are trying to determine the possibility of having Feed Conversion Ratio that is less than 1:1 in commercially reared broilers. Additionally, because most of the components of the chickens body are yet to be achieved through direct genetic selection, more researches tailored towards improving selection lines in order to suit new markets are required.

Although per capita broiler meat consumption is still highest in various developed westernized countries, more poultry production are shifting to various emerging markets, such Asian regions which are seeing the fastest poultry production growth rate. Considering the high cost of poultry production and other resources (land, energy, and poultry feeds), there is the possibility that various commercially reared broiler and egg laying chickens will go on been produced under less than the required conditions for the foreseeable future. Even in those already developed countries when poultry production is practised, the exorbitantly rising cost of poultry production have made most of the end customers to opt for previously strict management and poultry dietary practices in order to remain in the business. This has actually led to adoption of significant genetic technologies in the development of ideal selection methods that result in high performing and more healthy poultry strains.

THE ROLE OF GENOMICS IN POULTRY MEAT INDUSTRY

There have been several publications on development of full chicken genome in poultry industry, and this is actually an achievement towards incorporating genomic technologies in the production of commercial poultry breeds. Currently, most key poultry breeding companies are heavily investing in researches involving genomic technologies. These researches are tailored towards enhancing various effects and functions of genes in poultry breeding and production. For instance, geneticists have actually employed transgenic, which involves moving of genes from one breed of poultry to another artificially, in enhancing productivity in poultry industry.

Based on genomic researches and studies, breeders are now able to effectively and efficiently select breeds of high economic significant in poultry production industry. They are able to acquire traits that are beneficial in enhancing sustainability in poultry breeding structure.

GENETIC SELECTION AND BREEDING PYRAMID

Modern broiler breeding and production involves selecting and breeding pure breed lines with an aim to obtain desirable traits. Commercial broiler chickens are usually crossbred through three to four lines. The entire process usually starts by genetic selection of pure broiler breeds in pedigree breeding companies based on the traits and general performance of the birds. Once the breeds are genetically selected, they are moved to multiplication companies in order to be multiplied and crossbred. Those birds that exhibit various abnormalities are rejected. This breeding structure is commonly known as "broiler breeding pyramid."

Most commercially produced meat type chickens (broiler) are usually crossbred using at least three to four breeding lines. These breeding lines and the process of cross-breeding them are usually chosen and evaluated very carefully based on the ability to achieve the desired market demands. As mentioned earlier, the breeding structure that is commonly used in production of numerous crossbred meat type chickens is referred to as the breeding pyramid. Based on this breeding pyramid, the genetically selected birds are usually small in number and are found at the top of the pyramid. On the other hand, the number of crossbred broiler birds through three to four multiplication are usually numerous and are found at the bottom of the pyramid.

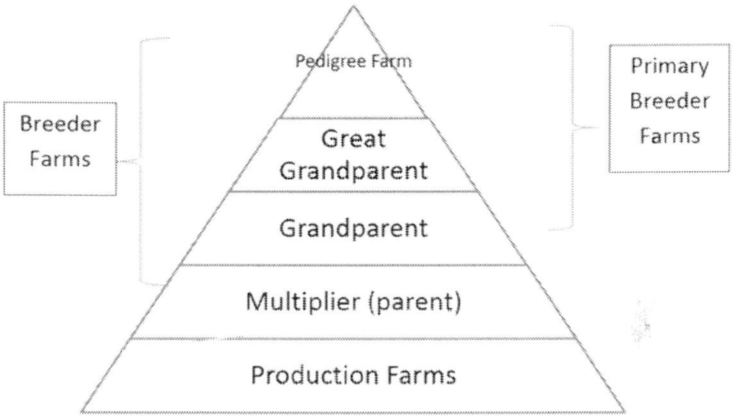

Figure 9.1: Structure and flow of genetics in the poultry industry.

Source: *http://aem.asm.org/content/72/5/3441/F4.expansion.html*

About 7,500 million meat type chickens are usually produced annually in European Union. To have such as large crossbred stock (7,500 million), over 60 million pure breeds are required. These broiler birds are usually produced by different breeding companies. Each company usually has its own pure breeding lines (maternal and paternal breeding lines), which are referred to as pedigree breeds/stocks. Each breeding company house its pedigree stocks in various geographically spread breeding sites that are mainly constructed for genetic selection purposes. Using these pedigree stocks, breeding companies produce grandparent and great grandparent birds through multiplication process. The grandparent produced stocks are sold. When breeding, most companies usually prevent exchanging their pure breeds with other breeds from other companies. This is to ensure that each purebred broiler line maintain the requirements for the market, which are ultimately dependent on the demand of various customers, retailers, and consumers.

When breeding, most companies usually ensure that subjective process of breeding goal is achieved. Additionally, all economic, societal, animal welfare, and other marketing aspects are also taken into account when breeding broiler chickens. Every breeding company usually has a specific type of commercial crossbred broiler breed or product that dominates in the local and international market.

During breeding process of the pedigree stocks, key biological traits of stocks are recorded at highly bio-secured breeding sites. Additionally, characteristics of various pedigree relatives are also recorded in non-bio-secured environments. This is usually to test the challenges experienced by the relative breed and the robust performance of highly bio-secured pedigree stocks. The values or traits observed on the pedigree stocks and on their relatives reared in different environment (bio-secured and none bio-secured) are recorded in a selection index which usually have determining factors for all breeding traits. Birds used for breeding (pedigree stocks) are usually ranked based on their selection index value.

After thorough evaluation on the traits of the bred stocks, a genetic selection process is performed to identify and select the best birds that should be used as the parents for the next generation. Genetic traits or relatedness of the selected birds should be as low as possible compared to those of the pedigree stocks. Additionally, their average selection index values should be higher than those of the pedigree stocks. The aid on the use of this selection is to maximize the genetic gain of the birds and to minimize chances of inbreeding in specific pure breeding lines. After selection is complete in pedigree stocks, the next step usually involves multiplication of those birds selected. There are three processes involved in multiplication of these birds:

- Pure-line multiplication
- Crossing pure lines in order to obtain crossbred parent stock
- Crossing crossbred parent birds in order to obtain commercial broiler birds

During multiplication process, all birds exhibiting certain abnormalities are usually culled or rejected in order to maintain quality of the final broiler products.

GENETIC BREEDING GOALS FOR COMMERCIAL MEAT PRODUCTION

The market requirement for commercial broiler chicken is quite different from those of the commercial turkey meat production. Different genetic selection methods are used for various meat reared poultry birds. Although market requirements are rapidly changing, meat type chicken industries are mainly focusing on live-market weight of the birds more than any other traits. This means that the chicken raised for the production of meat must fall within a certain range of market weight. However, those birds that fall outside these weight ranges are usually regarded as inefficient and may lead to some economic losses.

Turkey meat producing industries on the other hand primarily focus on how certain parts of the birds can be restructured into an improved final meat product. Weight of the birds is usually a main concern because most turkey breeders and producers usually market their birds at certain fixed age with substantial variations on the weight of the live birds throughout the whole year. Broiler producers also market their birds at an earlier point in their development (30% and 60% of their sexual maturity for the broiler and turkeys, respectively) than the turkeys. Therefore, there are slight differences when it comes to genetic selection of broiler methods and selection of the turkeys. Selection of broilers usually occurs much earlier as compared to selection of the turkeys. This difference is seen because of the elusive differences in production goals for turkeys and broilers.

GENETIC SELECTION FOR GROWTH RATE

Since the beginning of commercial broiler production in 1950s, growth rate of the birds was actually the first characteristic that receives most of the attention in meat producing companies. The main reason why most poultry meat producing companies focussed their attention on the growth rate is due to its economic significance and the trait can easily be manipulated and improved. Growth rate in poultry can be enhanced genetically through the use of genetic mass selection. Genetically, the process of improving the birds based on their growth rate involves mass enhancement, synchronous differentiation, and enhancing tissues maturation. Consequently, most results obtained after genetic selection is completed are usually influenced by specific strategies used such as primary age selection, selection intensity, selection based on correlated traits of the breeds, and various environmental factors (including nutritional aspects) under which selection is conducted.

There are numerous issues related to rapid growth rate selection of broilers. These issues include negative impact of the reproductive performance, issues on increased fat content in carcases, abnormalities on the skeletal part of the broiler birds, and other health conditions such as ascites resulting from physiological equilibrium disruption among other issues. Studies have shown that a number of these broiler issues can actually be ameliorated partially through the use of improved selection schemes. However, the complexity of these issues may actually complicate the entire process of mass selection for growth rate if they are improved.

Currently, there are different strategies been used by various commercial breeding companies for growth rates. Most of the programs used in selecting breeding birds are focussed on either the body weight of the birds or commercial age of the birds. Most companies are preselecting their breeding birds that are comparatively immature before finally selecting the birds at commercially appropriate age. Environment in which selection processes are performed usually vary from ideal environment which enables the birds to fully express their genetic growth rate potentials to the simulated environment with less influencers. Diseases are usually excluded in environment for commercial selection of the breeding stocks, but permits response evaluation to other social stressors, climatic conditions, density, and other physiological stressors.

Mass breed selection is commonly practised; however, different breeding companies are presently employing other complex selection strategies for estimating breeding value. These strategies include single-trait indices selection, multi-trait indices selection, family selection, and Best Linear Unbiased Prediction (BLUP).

SKELETAL ISSUES RELATED TO GENETIC SELECTION FOR INCREASE POULTRY MEAT PRODUCTION

Genetic selection is a modern method that is commonly used by most poultry meat producers in selection for improved meat and eggs production. Although this strategy is beneficial to the producers, it has potentially affected the welfare of different breeds used in poultry industries. Some of the common issues include negative effects on skeletal integrity of broiler chickens and mineralization issues in egg laying chickens; growth and strength of ligaments, tendons and bones; and musculoskeletal infections development in broiler birds among other skeletal complications.

Since poultry genetic breeding was established in 1950s in order to increase productivity and enhance profitability, little emphases have be put to alleviate various skeletal welfare issues commonly associated with genetic selection for improved production in poultry. In egg laying breeds, genetic selection has been used to enhance eggs production resulting to low body weight of the birds that produce numerous numbers of eggs with less poultry feeds intake. Annually, the quantity of the used calcium in formation of egg shells is usually 20 times the total content of their body. Most laying birds use calcium in their bone for the formation of egg shells, and this usually result to suppression on formation of the bone due to loss of bone mineral thus resulting to skeletal problems (a condition commonly referred to as osteoporosis). Some of the common skeletal problems usually seen in layers include paralysis, fracture of bones, poor bone quality, and death may result if the problem is not handled amicably.

SPECIFIC GENETIC LEG DISORDERS IN POULTRY INDUSTRY

Different studies have demonstrated the link between genetic selection of certain poultry breeds and heritable nature of various disorders of the legs. Some of these disorders include twisted leg disorders, and tibial dyschondroplasia.

Twisted Leg Disorders

This is a common condition mainly found in meat type reared chickens that have genetically been selected for increase growth rate. Twisted leg disorders in chicken can be differentiated into rotational defects or valgus-varus commonly seen in live chickens. To measure the intensity of rotation or bending of the tibia, dissection can be performed on the long bones of the chickens in order to obtain detail information on the nature of deformities, including details on the proximal tibia plantar bending which is often thought to be as a result of tibial dyschondroplasia. Referring to an experiment performed in 1984 by Leenstra and colleagues to study occurrence of twisted leg disorders in different poultry breeds, three experimental breeding lines were genetically selected and put to test. The first line involved breeds for high body growth rate-six weeks which was denoted by letter (C), the second breeding line involved a full-sib genetic selection for the purposes of induced incidence of twisted legs denoted by letter (T), and the third breeding line was breeds genetically selected for reduced incidence of twisted legs combined with breeds genetically selected for high body growth rate-six weeks which was denoted by letter (CT). These three breeding lines revealed strong genetic relationships on the occurrence of twisted leg disorders as shown in the figure table below.

Table 9.1: The effects on occurrence of twisted leg disorders and body weight from three genetically selected breeding lines

Occurrence of Twisted Leg Disorders (%)						
	In cages		**On litter**		**Body weight at six weeks (g)**	
Breeding Line	Male	Female	Male	Female	Male	Female
C	49	28	18	12	1969	1708
T	12	7	5	2	1808	1555
CT	27	12	8	5	1941	1677

Source: *http://www.scielo.br/scielo.php?script=sci_arttext&pid=S1516-35982016000700372*

Studies have shown that most twisted leg disorders are influenced by incomplete expression of autosomal recessive genes. These genes are usually responsible for key genetic variations on the occurrence of twisted leg disorders in broiler chickens and other poultry birds in the farm.

Tibial Dyschondroplasia (TD)

Since evolution of meat production industry in nineteenth century, poultry meat industry has always been dedicated towards raising the demand for the processed and portioned poultry meat products. One of the main and valuable parts in the carcase of a chicken is the breast muscle. Most breeders are putting more attention in this part because of its value in production of chicken meat. Studies have revealed that breast muscle meat have a high level of genetic heritability that range from about 0.45 to about 0.6. For instance, it has been revealed that the modern genetically selected broiler strains, raised under proper hygienic condition are able to produce about 19% of breast meat when they are compared to the same broiler strains reared like 30 years ago, when the production of breast meat was only 10 to 12%.

Regardless of enhanced breast meat production on modern broiler chickens, most breeders still do not understand the consequences resulting from genetic selection on breast meat quality. Studies also show that more broiler breeders are currently focusing on the genetic selection of strains based on the ability to reduce fatness. This is done because fattening during the period of rearing the birds usually raises broiler product final cost. In other words, fat broiler birds usually have high mortality, higher feed ratio: weight gain and also reduced value of the carcass at the meat processing plant. Additional, most customers prefer chicken meat because of its leanness, therefore more broiler breeders are emphasizing on the importance of producing less fat meat in order to meet the demand of most customers. Various genetic poultry strategies have been adopted with an aim of reducing the content of fat in poultry birds raised for meat production.

The occurrence of tibial dyschondroplasia has been compared with different genetic work adapted to growth rate, high level of ambient temperature; high poultry feeds efficiency, and high broiler meat production. It has been revealed that broiler meat production usually go down when the birds are raised or reared in environment with high levels of temperature. For this reason therefore, modern meat chicken are genetically modified to reduce feathering with an aim of help the birds to efficiently dissipate their internal heat.

BEHAVIOURAL ISSUES RELATED TO GENETIC SELECTION FOR IMPROVED PRODUCTION IN BROILERS

Domestication of modern poultry flocks was mainly done because of various favourable traits exhibited by the wild ancestors. Some of these traits include sexual behaviour exhibited by male chickens been dominant over the females and mating signals, hierarchical group structure, early parent-offspring attachment, adaptability to different environmental conditions, and non-specific to nutritional requirement among others. Some of these traits are used in the intensive poultry farming today, while others are not important especially when it comes to selection of breeding birds for meat production. This is because most poultry breeders have genetically altered the behavioural traits of the modern chickens in different ways. First there is the adoption of fast growth rate technologies, feed conversion enhancements, breeding strategies such as the use of artificial insemination, poultry housing system such as the used of caged system and deep litter system among others, artificial incubation strategies, and the use of environmentally controlled housing systems.

Several years of continuous genetic selection while emphasizing on improved production have not linked the key changes obtained in behavioural characteristics needed for the birds to adapt to various environmental conditions. This has actually led to various health issues and maladaptive behaviours. Some managerial activities such as low light intensity and beak trimming have occasionally be used to mask various behavioural problems that result from genetic selection of breed for high production capacity. Some of the significant behavioural issues associated with genetic selection in broiler chickens for fast growth rate include skeletal disorders and severe feed restriction usually imposed on broiler breeder chickens.

Modern broiler chickens are usually reared in social context and this usually makes it difficult for them to establish ideal social relations. This usually result to extreme aggression causing some injuries, which in turn lead to high mortality due to cannibalism and exposure to other bacterial infections. Intense genetic selection pressure is been applied for improved production in broiler meat thus resulting to various correlated impacts on the social behaviour of the broiler chickens. Studies have shown that those genetically selected broiler birds either for early maturity or for high meat production are socially dominant than the unselected birds.

Most broiler chickens are genetically reared in environment with low external stimulation. These birds usually find it hard to cope with any subsequent change in environment condition. Most of them usually overreact to different apparently harmless stimuli. Inappropriate response to fear usually causes pain, injuries, and death to many broiler birds. Fearfulness in broiler chickens usually leads to low production. Expression to fear in broiler chickens is positively correlated with the level of restriction usually imposed to them. This has always been considered as one of the key factor of poor welfare in broiler meat producing chickens.

Studies have also shown that most broiler breeder chickens usually restricted to feed are constantly hungry. Most of these breeder chickens are usually restricted to feed genetically in order to maintain high levels of productivity in their progenies. These restricted birds usually maintain their reproductive competence even if they are severely restricted to feed. Artificial insemination can be used in order to minimize the need for restricting breeder chickens; however, it may lead to other different welfare and behavioural issues.

ETHICAL CONSIDERATIONS AND THE FUTURISTIC FATE OF BROILER CHICKENS

Research findings have shown that the welfare of an animal is highly influenced by various environmental factor, its genetic traits, and interactions between various genetic traits and environment factors. This means that a broiler chicken is able to adapt to a new environment and various factors of that environment may influence the physiological and behaviour plasticity of the birds. Adaptation outcomes may result to improvement on the welfare of the broiler chickens by enhancing the birds' fitness for a certain environment, thus leading to improved contentment. Genetic poultry selection for specific phenotypic traits related to certain behavioural or psychological display has actually become a key tool in improving production in broilers. According to studies, it has been revealed that both productivity and welfare of the broiler chickens can actually be enhanced through the use of genetic selection methods.

For instance, with the use of genomic sequences, breeders can actually select fast growing broiler chickens that have the potential to enhance production and at the same time maintain optimal welfare of the chickens. This may help in alleviating stress related problems and various broiler related health conditions. Some of the currently used methods that can be used in selecting broilers chicken for improved production and optimal welfare include (1) genomic poultry selection strategies and (2) multi-trait and multi-level poultry selection strategies.

There have been several ethical protests fighting for the welfare of genetically engineered broiler chickens and other animals in the meat production industries. The primary focus has always been geared towards broiler integrity violation. Additionally, attention has always been given on the individual suffering of the broiler chickens and other moral repulsion against defining broiler chickens as patentable animals.

Broiler chickens used in genetic selection are usually degraded through be used as genetic sources, DNA sequences, production traits, model system, and body part for improved production. This means that different broiler chickens used in genetic selection (pedigree) are no longer considered as broiler chicken, rather than pieces for improved meat production.

Modern meat type chickens (broiler) have been genetically selected for remarkably high growth rate. Because of growing abnormally and so fast, most of them spend much of their time lying down on the floor while others suffer from lameness. Fast growth rate on broiler also put more strain on the heart and lungs of the chickens.

Most genetically selected and reared breeder chickens are usually housed in factory farms and forced to stay there for a period not less than one year. Because these birds are kept alive for a period longer than the ordinary broiler chickens, a big number of them usually end up developing various organ failures. Most of them end up dying as they grow bigger and bigger due to their bodies that have been genetically manipulated. In attempt to prevent them from growing, most breeding farmers usually limit feeds for these birds drastically, thus keeping these breeder chickens in a persistent state of frustration and hunger. After a long period of confinement and deprivation, these breeding broiler birds usually look so worn out with emaciated bodies. Because they are usually genetically unenviable, they are loaded on trucks and sent for slaughtering.

CONCLUSION

Different poultry meat production companies practice intense genetic selection for various economically significant characteristics such as growth rate, high body weight, efficient feed conversion and most importantly for the traits related to carcass-processing that have positively contributed to improved efficiency and productivity in broiler production industry. Although most of these enhancements on broiler meat production have contributed to different unintended results such as -decreased performance in reproductive of the birds, ascites, skeletal abnormalities, and abnormal increment in the abdominal fat content- most of them are usually ameliorable through practising proper nutrition and management, and genetic selection will actually go on been the key driver for poultry improvement and productivity.

CHAPTER 10

GENETIC IMPROVEMENT IN LAYERS FOR EGG PRODUCTION AND QUALITY

INTRODUCTION

For the last few decades, there has been significant improvement in commercially reared poultry for egg production in terms of enhanced production and improved quality of the eggs. Most egg layers breeding companies have been focusing on how the total annual number of eggs per bird can be maximized, putting into consideration other factors such as the welfare of the chickens and health. For instance, breeding companies have managed to genetically improve the total number of eggs yearly laid at the 80 weeks of age from between two and three eggs and the quality of egg shell from between 20 and 100 grams. These production improvements have actually been attributed by two key factors, which are genetics selection and environmental factors (such as feeding, housing system, hygienic status, and other zootechnical factors).

Breeders are concentrating more on the use phenotype idea to help in improving the quality of eggs and overall production. However, collection of phenotypes in poultry birds is a tricky method and should be done accurately in order to have the best results. The quality of eggs is important because it influence the final breeding goal such as breaking strength and weight. The idea of collecting phenotype in egg laying breeds has actually help the breeding companies in determining other traits like feeding behaviours of the birds (feed intake and feed efficiency). The environment in which these phenotypes are collected is also significant. Pure breeding lines birds are usually housed in well governed house with up-to-date biosecurity standards in order to enhance the quality of products that are free from various diseases to meet the required international veterinarian demands. In other words, this means that pure breeding lines are usually done in perfect and ideal environmental conditions.

For international egg layer breeding companies, ideal environment alone is not enough. It is required that breeding lines to be done both in ideal environments and in practical circumstance (field test). The field test involves breeding lines that perform in different sanitary condition, housing system, feed and water quality, and different temperature and humidity. All the characteristics on the field tested birds should be measured against those characteristics obtained from the pure breeding lines reared in ideal environmental conditions. This actually helps the breeders in determining the robustness and survival of the ideal breeds for breeding purpose. All these birds are wing linked and branded in the poultry database using information on pedigree. Another most important advantage for using field test is because it helps egg layer breeders in determining genetic potential such as the total number of eggs and quality of eggs of the pure line males. The males usually do not produce eggs but, like the female breeds, they are genetically potential in influencing the quality and total number of eggs.

GENETIC VARIABILITY OF EGG QUALITY AND PROSPECTS FOR SELECTION

The main purposes for egg production include reproduction (hatching), for direct consumption, and for further processing into other products of eggs. Based on these three channels for egg marketing in commercial egg breeding and production programs, genetic selection play a very crucial role in determining these market aspects or channels. For instance, hatchability is usually not considered as the main purpose, but the quality of the internal content of the egg is primarily the most relevant area among the three channels. Contrary to various egg processors, stability of the egg shell is of greatest importance to both the table eggs and breeder producers.

Therefore, improving the quality of egg has always been the main aspects of various breeding companies for egg-laying chickens. Genomic and genetic studies have actually help in identifying new methods of improving the quality of eggs such as the use of high density genotyping in order to allow genome wide selection. Genetic and genomic selection have enable development in different egg traits measurement such as cuticle and pigmentation, quality of the shell, quality of albumen, and egg yolk quality.

Cuticle and Pigmentation of the Eggs

Cuticle is the structure that actually helps in preventing different microbes from entering into an egg. Studies have shown that the amount and quality of egg cuticle is mainly influence by different genetic factors of the eggs. Genetic selection of breed hens for egg laying that possess the best cuticle coverage is significant especially in prevention of bacteria penetration into the eggs. This is a crucial aspect especially when it comes to maintain good consumer safety, reduction of egg spoilage, and lowering chances of transferring microorganisms from generation to generation.

Eggs pigmentation or colour, although it is not often considered as an important aspect in prevention of egg damage or safety of egg, it does play a crucial role in marketing of the egg. Studies have shown that different markets have strong preferences on the colour of eggs. Although the pathways used in pigmentation of the egg shells are usually comparatively simple, little has been done in an attempt to enhance variation in pigmentation of eggs or even showing the difference between the white eggs and brown eggs. The only exception is research conducted in identifying 'O' loci in chickens that produce blue eggs as a result of secretion of biliverdin in their bodies. Genetically, this has been recorded to chromosomes and different markers that are close to the loci have also been mapped.

Those chickens that produce blue eggs usually have got high expression ofheme oxidase, which usually helps in formation ofbiliverdin. Although most heme oxidase genes are located on chromosomes, they are usually about 13Mb away from the loci recorded to be responsible for the phenotype. Studies have actually indicated that there is no close link between the colours of the shell with various markers located at heme oxidase loci. It is usually assumed that heme oxidase expression is altered by the loci transcription factor, but the responsible gene has not yet been identified. Therefore, given the high trait heritability in poultry, the genes responsible for colour alteration might be identified easily. This will actually be a significant step to most breeders in egg production. They will be able to create breeding lines with a variety of colour traits which will in turn allow hybrids production appropriate for different markets.

Egg Shell Quality

Different studies have been done to find out how eggshell quality can actually be improved. However, the complexity traits of the eggshell and its sophisticated structure have always posed a challenge in meeting the right quality. In other words, different genetic factors influence the traits expression on the quality of eggshell. Application of genomic on the glands of eggshell has made it possible in identification of different variation in quality of egg shell. Researchers have also combined genomic and genetic resources in determining the how the quality of eggshell can be enhanced. Different association genome studies have also been published in order to give various companies that are undertaking egg production renewed impetus on how to improve egg production and to enhance quality of the eggs.

Maintaining the quality of eggshell as an aspect of egg production has been a key goal of genetic selection. The main idea of improving quality of shell has always been to prevent damage especially during packaging and transportation of the eggs to the end consumers. The success on improving eggshell based on genetic selection has been evaluated through experiments to observe the differences between quality of eggs in traditionally reared breeds, which are comparatively poor egg layers and pure genetically selected commercial egg laying breeds. Some of the major strategies employed to determine this success include determining shell weight, determining specific gravity, which helps in estimating the weight of shell, and quasi static compression (such as deformation and puncture or breaking strength).

A few cases have been reported of using the number of cracks on eggs laid by individual birds, presumably because of various predisposition environmental factors that results to variations in collected data. Variations on data collected occur because insults that egg receives from the time it was laid until it reaches the end consumer are usually uncontrollable. However, using heritability that range from 0.50 to 0.62 recorded when a threshold strategy on frequency of egg crack may help to give a good method to actually be able to capture this information, especially if it is validated in various commercial egg breeds.

Another factor that has receive great focus from the geneticists in trying to maintain the quality of egg shell is dynamic stiffness, which is also influenced by heritability of the breed. Stiffness of egg shell was actually assessed and evaluated at the University of Leuven with the help of eggs resonant frequency in the year 1999. From this study, it was revealed that genetic correlation between the breaking strength and stiffness of the egg was (0.49). Genetic selection has helps in improving the stiffness of the eggshell in order to increase shell-life of the eggs until is reach the end consumers.

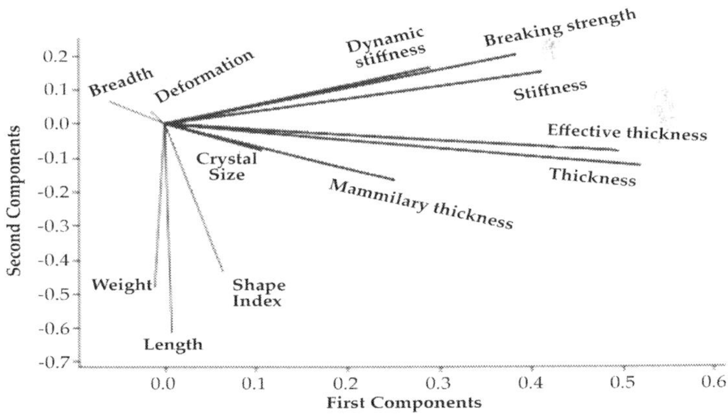

Figure 10.1: First and second principal elements of eggshell quality.

Source: *https://www.slideshare.net/abohemeedaly/poultry-feeding*

Based on the principle elements in the figure below that determine the quality of egg shell, it can be deduced that the thickness of the shell, the crystal size of the eggshell, breaking strength of the eggshell, and dynamic stiffness of the shell are the primary components that influence the quality of the egg shell. Other traits that determine quality of the egg shell include the weight and shape of the shell. Most of the information on these components can be acquired through genetic correlation between the traits.

Albumen

Just like in the case of eggshell quality, genetic selection has actually help in enhancing egg production and marketing through increasing the quality of egg white. The quality of egg albumen is primarily estimated through the use of Haugh Units or albumen height which have severally indicate high heritability that range from 0.2 to 0.5. However, the amount of protein is of more practical significant in egg production and processing industries. Genetically, albumen height or Haugh units usually have poor correlation with the content of protein in an egg. In other words, this means that albumen height or Haugh unit is mainly correlated with fresh egg appearance than the nutritional composition of the eggs.

Geneticists are researching on viable methods to actually estimate the content of protein in eggs, which will not only help in enhancing the appearance of fresh eggs but also the processing and nutritional qualities of the eggs. Some of the methods been used currently such as low-resolution proton nuclear magnetic resonance and near-infrared transmission spectroscopy are not up-to-date and have occasionally been disappointing. Although eggs have numerous proteins in the albumen, there are very few that contribute to the aspects of egg quality. However, through genetic research, geneticists have achieved a lot in identification of ovotransferrin which forms a perfect starting point to enhancing egg production and quality of eggs protein.

Another trait that affects the quality of egg albumen is presence of blood spots and meat in the albumen. Geneticists are doing all they can to actually find out what causes this phenomenon. However, studies have shown that heritability of this phenomenon is usually low (0 to about 0.18) especially in brown egg laying chickens. This is because of the similarities on the eggshell pigmentation and both the blood and meat spot in the albumen.

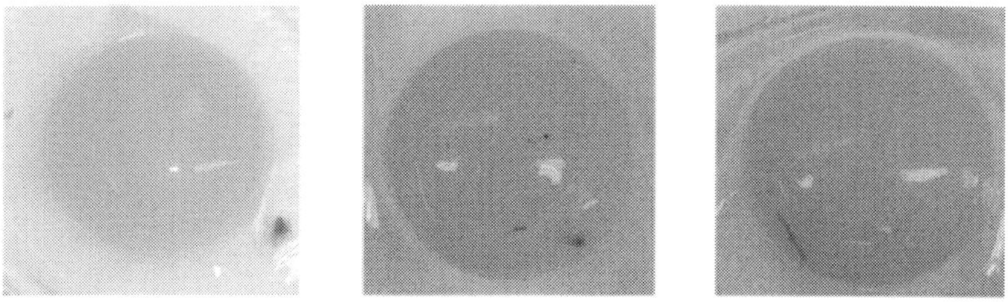

Figure 10.2: Appearance of different size of blood and meat inclusions in eggs.

Source: *https://www.backyardchickens.com/articles/common-egg-quality-problems.65923/*

The antimicrobial potentiality of egg white is a significant aspect in production of eggs. Although most customers are not likely to bother about this aspect when purchasing their eggs, genetic selection would probably help in reducing the risk to the end consumers and also reduce transmission of the microorganisms from one breeding generation to another. A good example is transmission of salmonella, which bear modest and useful heritability traits that can be reproduced in subsequent breeding generations.

Egg Yolk

Eggs nutritional value is largely determined by the egg yolk content. Egg yolk usually account for about 30% of the total content and weight of an egg. Additionally, about 50% is accounted by the dry matter. This therefore means that if the dry matter of the egg is to be changed, this would automatically lead in alteration of the albumen to egg yolk ratio.

Based on studies, it is practically possible to determine genetic selection response in the proportion of egg yolk in various breeding chickens. The estimated value for heritability is relatively high (0.38). The proportion of egg yolk can be obtain by getting the ratio of the yolk weight against the weight of the egg. Therefore, there is a positive genetic correlation between the weights of the eggs and the weight of the yolk, and determining this correlation accurately would help in improving hatchability of the large eggs.

Genetic selection has played an important role in the proportion of egg yolk and thus influencing the dry matter of the egg. This is a trait with substance economic significance in egg production and processing industry. Egg yolk is also a primary source of nutrient for the development of the embryo during incubation period. This means that the composition and viability of the chicks body is highly determined by content of egg yolk.

Genetic measures have also been employed to improve the quality of egg yolk, including programs to help in eradicating various defects in genes that are responsible for the fishy taint in the egg yolk.

The proportion of the egg yolk is therefore not only essential aspect in terms of egg production performance, but it is also an economically significant aspect for egg processors and should genetically and practically be enhanced in commercial egg laying and breeding programs.

GENETIC MECHANISM WHICH CONTROLS VARIOUS FACTORS AFFECTING HATCHABILITY OF A FERTILE EGG

Hatchability in poultry is a complex trait that is occasionally influenced by the age of the birds. It is comprised of other different sub-traits which are usually susceptible or influenced by both environmental and genetic factors that arise from different sources. There have been several studies conducted to examine the genetic factors influencing production of eggs, quality and fertility of the eggs, and how these genetic factors correlate with hatchability traits.

It has also been revealed that both females and males genetic elements are usually significant in determining hatchability traits. This is because of the importance of these elements or components on fertility and early mortality rate of the embryos. The female that produces the eggs is usually the main source of various genetic variations that influence hatchability of fertile eggs. The effects of genetics (both maternal and paternal) usually vary based on the age of the flocks.

Biologically, the primary role or purpose of eggs is to offer a suitable environment for embryo formation and development, usually outside the body of a bird. The final outcome for a single egg is usually a simple binary response. This means an egg may either hatch into a new chick or it does not. However, the process of successfully forming and producing a chick is usually complex and requires fulfilment of a number of conditions, which include;

An egg has to be produced by a bird (a hen in this case). Hatchability of this laid or produced egg mainly depend on environmental and genetic elements influencing the ability of a female bird to lay the egg.

- The egg must be fertile. This means that for hatching to take place successfully, mating between a male bird and a female bird should take place or alternatively artificial insemination should be practised. In this case viability for embryo formation in the process is highly influenced by both genetic and environmental factors from both male and female birds.
- The quality of egg laid should be optimal in order to offer an appropriate environment for the embryo to form and develop. The quality of egg is determined by the genotype of the birds and other external factors surrounding the egg during storage and incubation period.
- The final condition involves hatching of the egg into a chick. The ability for the embryo or fully developed chick to break out of the egg shell during hatching period is usually influenced by the embryos genotype, the internal surrounding of the egg, and external surroundings to which the egg was exposed.

Production of an Egg

Production of eggs is usually evaluated based on cumulative eggs record from the time a hen start to lay eggs up to a certain age. On limitation to this approach is the fact that most breeders ignore that most eggs are produced or laid over a long period of time. During this period, the bird (hen) producing the eggs and her mate are usually predisposed to various environmental and genetic factors. Studies have actually revealed that egg hatchability usually differ between the early production stages and late production stages. In other words, there is always a significant variation in the heterotic, Z-chromosome, and additive genetic effects of laid egg based on age of the birds.

Different models have been put into practise to evaluate and assess egg production in order to analyse various changes in variance aspects as the egg laying bird age. For instance, breeders have used cumulative egg production on a monthly basis to analyse various traits. Fixed regression model has also helped breeders to analyse repeated measures of similar traits over a period of time. When using this model, it is advisable to always consider the flock production curve but assume that all genetic factors are constant.

However, results from different studies conducted in poultry production reveal that egg production in early stage and in late stage of production are usually different traits genetically. Although egg production heritability is relatively low to moderate, there is huge genetic variation in egg production of modern breeds despite numerous genetic selections from one generation to the other.

Fertility

Egg fertility is usually influenced by various factors that originate from the egg laying birds. These factors include ability to successfully mate, sperm storage, egg cell ovulation, and suitability of the surrounding aspects for the formation and development of the embryo. Fertility of the egg is also influenced by the quality and quantity of deposited semen. Most of these factors are subjective to the age of the bird. In most of the studies, fertility of the bird is usually treated exclusively as an aspect of the female bird producing the eggs (female fertility) or the male bird that mate with female bird for egg fertilization. There are different study models that have been employed to determine the fertility in poultry. Maternal effects model-for putative gene; repeatability model-for random non-genetic effects on male; and threshold model-for assessing simultaneously effects of both female and male genetics.

Quality of Eggs

Genetically, quality of an egg can be discussed based on two major categories, which are internal quality and external quality. The external egg quality is usually influenced by various genetic features of an egg such as shape of the egg, the size of an egg, strength of the eggshell, thickness of the eggshell, and general structure of the egg among other aspects. On the other hand, internal quality of an egg is usually measured based on the genetic quality of egg white (albumen) which is denoted by the Haugh Units (HU). Another method involves determining the relative size of various components. Additionally, determining the integrity of the shell membrane can also help in determining internal quality of an egg. Different studies have been conducted to assess the quality of egg in chickens as well as various micro environmental factors that influence quality of the egg during storage, during embryonic formation stages, and how these factors influence hatchability of the fertile eggs. For instance, studies have shown that the value of Haugh Units (HU) usually decline while the PH value of albumen increase during the storage period. It has also been revealed that moulting of the female chickens actually helps in enhancing hatchability of the eggs especially if the eggs were stored for a long period of time. This is because moulting genetically improves HU of the stored eggs.

Embryo Survival

The mortality of embryo is never distributed evenly in the incubation period. The high cases of embryo death are usually recorded during the first and third trimester of incubation period. Survival probability of an embryo can actually be described as its genotypic function which is influenced by the genes obtained from the parents, and the environment of the egg which generally depends on the parents. Early embryo death is usually attributed to aberrations of chromosomes and lethal genes received from the parent birds (both female and male).

Studies have also revealed that susceptibility to embryonic mortality is heritable and usually decrease as the embryo growth and development progresses (from about 0.09 for early mortality to about 0.05 for late mortality based on female components) and (from about 0.25 for early mortality to about 0.18 for late mortality based on the male components).

In summary, hatchability in chicken is described as a function of both the male and female birds. This is because of crucial effects contributed by both paternal and maternal aspects on fertility and survival rate of embryos. However, there are several genetic variations that occur in hatchability of fertile eggs laid by a female bird. These variations usually depend on the quality of laid eggs (both internal quality and external quality), and affects formation and development of embryo during incubation period and chick emergence from the egg when hatching.

GENETICS OF BROODINESS IN EGG LAYING BIRDS

Broodiness is a behavioural response exhibited by most egg laying chickens. Most commercially reared chickens that have been genetically selected usually do not go broody. Because of the fundamental role attached to broodiness in poultry reproductive system, different geneticists and genomics have taken part in investigating these behavioural traits in order to enhance egg production. Genetic selection against broodiness behaviours in poultry has successful been achieved in White Leghorn breeds that has led to completely loss their avian ability and parental behaviour (egg incubation). Studies have actually revealed that genetic in broodiness is a polygenic trait which is related to key sex-linked effects. There are also some proves that broodiness in egg laying chickens is influenced by key autosomal genes that contribute to behaviour expression in poultry. This means that if there are no broody genes on the Z chromosome, some of the autosomal genes might be responsible for incubation behaviour inhabitation in egg layers.

Reduction in broodiness has also been observed in turkey because of the correlated response to genetic selection in order to enhance eggs production. There has been a continuous inhibition of incubation behaviour in female turkeys as a strategic method of improving egg production performance. This has led to ovulation termination and ovarian regression.

Broodiness in poultry is usually influenced by various environmental conditions such as darkness, high temperature levels, presence of chicks, and removal of the eggs in the nest as they are laid. Studies have shown that these conditions may even evoke those breeds that do not go broody like White Leghorns. The differences of broodiness expression in various breeds reveal that this is a hereditary trait that can be influenced genetically.

Geneticists and other poultry breeding experts have shown that Z-linked genes are usually involved in inhibiting broodiness in egg layers. This is done through cross breeding between different egg breeding strains.

COMBINED EFFORTS OF NUTRITIONISTS AND GENETICISTS TO REDUCE FEED COST IN LAYERS

In poultry rearing, especially for the broiler and egg laying breeds, feeds usually account for about 75% of the total egg and meat production. This production element is getting a lot of attentions from both the breeders and producers especially where focus is put on profitability of a business. Based on different genetic studies that have been done on the nutrients requirements of different poultry species, geneticists and nutritionists are able to come up with ideal formulation for "least cost" egg laying feeds. This is done taking into account different prices of various components used to make the feeds and the production level of the birds. It has always been perceived that egg laying chickens that are fed in a well-balanced diet ration will consumed enough in order to satisfy their nutritional requirements for better egg production and general maintenance purposes. Geneticists and nutritionists could mainly focus on formulating poultry feeds that helps in meeting the nutritional needs at a lower cost of feed per tonne.

In the year 1970, when the prices of eggs were low, the prices of feeds were high. Because of this, various geneticists begun to look for ways of improving feed efficiency in egg laying birds. One way that was adopted was using intensive genetic selection of breeding stocks. Instead of just getting the weight of the egg layers to approximate maintenance needs of the birds, they measured feed intake of individual birds in order to evaluate the "residual" feed intake and "the income of the eggs based on the cost of the feeds". From record analysis made from the pedigreed egg laying chickens, it was confirmed that those egg producing chickens with comparable egg mass production and body weight differed more on feed consumption than previous breeds. Studies have shown that feed conversion efficiency in egg laying chickens has enhanced substantially since 1970s (from about three kilograms to about two kilograms feeds per a kilogram of egg mass).

Additionally, there has been a lot of information on various factors that influence consumption of feeds and efficiency of converting those feeds in poultry eggs and meats. These factors include housing temperatures, levels of water intake, feather coverage on the birds, changing nutrient needs and feeding structures with age, and time of the day. As shown in the figure below, it is clear that efficiency in feed intake and conversion rate in poultry for the last 50 years has continuously been enhanced, and this has always been influenced by different management systems.

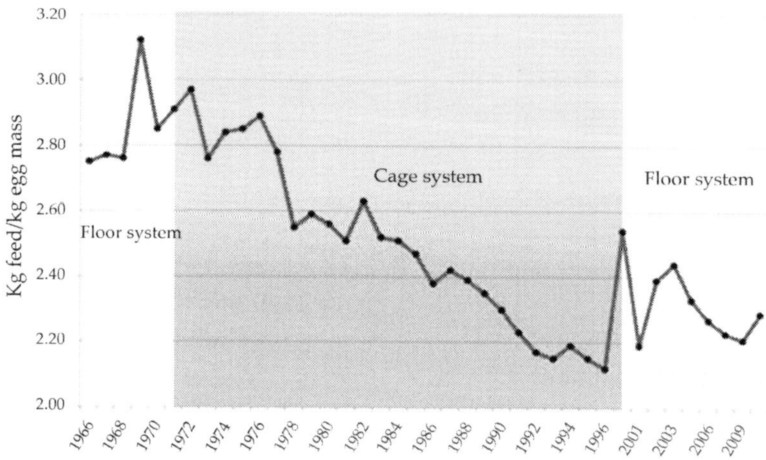

Figure 10.3: Effects of improved feed efficiency using genetic selection and proper management practices.

Source: *http://www.sciencedirect.com/science/article/pii/S0022030216301655*

ADVANCES IN GENETIC TECHNOLOGY FOR CONTINUED IMPROVEMENT IN EGG PRODUCTION

Most of the egg consumers are currently interested not only in the eggs quality but also other factors that promote health and welfare of the birds. These include sustainability and housing factors. On the other hand, the producers are mainly interested on how the birds can be improved genetically to produce more eggs at a competitive cost. These issues of production alternatives, the welfare of the birds, beak treatment, environmental footprint, and sustainability have contributed to several challenges in the husbandry and production of eggs.

Modern trend experiences in egg production industries suggest or indicate that most birds are biologically engineered in order to enhance their efficiencies in production of more and quality eggs than birds marketed a few decades ago. Studies have shown that today's egg layers consume less feeds, but produce more eggs than before. They are also very sensitive to marginal regimes of nutrients and other extreme fluctuations in environmental conditions. If these factors are combined, they would result to significant strain interactions with environment. Additionally, those regions that are currently facing extreme and hostile climatic conditions such as high temperature and humidity levels and challenging poultry diseases are the ones with the greatest potential for poultry production.

Genetic selection has therefore been directed in improving strains of egg laying birds in order to genetically make them resilient and robust. Genetics has also helped in enhancing the quantity, quality, and optimal egg size produced by individual birds. Apart from these traits, there are other elements of egg production that have been modified genetically such as the texture and strength of the eggshell for better storage and transportation, better internal egg quality (reduced meat spot, blood spots, and Haugh Units), improved feed conversion efficiency, high egg and bird survival, improved egg liquid content, improved albumen and egg yolk solid content levels, and other functional properties that have helped in meeting the rapid increasing producers and consumers demands.

Current Performance of Layer Strains

Maximum level in performance achievement in layers is usually determined by the genotype of the birds. Modern brown and white commercially reared egg layers have been modified genetically to enhance their potentiality to reach sexual maturity within a short period of time. Genetically, they are able to achieve about 50% Hen Day (HD) production within a period of 140 days of age. Optimal or peak Hen Day production is usually at 96–98% and they are capable of maintaining Hen Day production levels of above 90% for a period not less than 25 weeks. Egg layer birds that are reared for a period of up to 80 weeks of age have the potential of producing about 345 to about 364 eggs per bird with an average egg weight of about 60 to 65 grams. Based on these egg production oversights, the average feed consumption is usually about 22.0 to about 23.0 lbs per 100 hens on a daily basis. Feed efficiency on the other hand is usually estimated from about 2.0 to about 2.2 (g/g) while the survival is usually between 94 and 97%.

Egg layers reared up to a period not less than 90 weeks of age usually can produce about 398 to about 413 eggs per hen housed and the average egg weight is usually between 62–64 grams. The mass of the egg is usually between 55 lbs. to about 57 lbs, while the average feed consumption is usually estimated to about 22.7 lbs to about 23.5 lbs for 100 birds in a day. Feed efficiency is usually between 2.0 to 2.3 (g/g) while survival is about 94 to 96%. Genetic advancement in poultry diseases control, poultry nutrition, husbandry and management system practices have greatly influenced these achievements of high egg production and performance.

Genetic Improvement Methods for Continued Genetic Progress

Most eggs laying birds have for a long period of time encountered mystifying environment conditions in various poultry production industries. Most of these environmental challenging conditions influence the biological systems of different egg layers causing them unable to fully express their genetic potentiality in egg production. Geneticists have taken the initiatives of developing layer strains that are competitive enough and have superior biological traits and systems able to overcome various challenging environmental conditions. Modern strains are now able to fully express their genetic potentialities in the array of various economic characteristics that are necessary in egg production industries. Therefore, egg laying strains continue to receive genetic improvements in their biological systems.

Emphases have also been given in maintenance of the immune competence of the egg layers. This is significant for the overall health status of the birds, welfare, and productivity of the birds. Genetic selection has helped in improving the capacity of immune response resulting to egg producing birds that are resistant and robust to deal with various challenging environmental conditions.

Through the use of genetic selection strategies, more and more egg producers have continued breeding birds for extended period of time without moulting. Studies and researches have shown that interaction of gonadal, hypophyseal, and neuroendocrine responses responsible for various reproductive functions in the body of egg layers must be kept in synchrony with various behavioural responses such as moulting. However, this has been observed to deteriorate as the birds progressively age with time resulting to senescence. Using the modern genetic selection methods that helps in extending egg production cycle without moulting has enhanced these responses in commercially reared egg laying birds.

Advanced genomic studies have actually helped most geneticists in evaluating and estimating the values of breeds using information on DNA. This has actually assisted in improving egg production, particularly through obtaining strains with high survival rates and with the best natural immune defense. Using both quantitative genetics and genomic approaches for selecting the best egg laying strains, statistical strategies to help in improving model analyses through interaction between the environment and genotypes have been applied in different poultry breeding programs in order to enhance estimation accuracy of breeding value. Using pedigree testing in cross-breeding lines in various field conditions has also helped in improving estimation of breeding value by lowering the effects of interaction between the environment and genotype, thus improving progress in genetic selection.

Maintaining feed consumption is an essential aspect for egg producing chickens. This is essentially crucial under certain environmental conditions such as excess heat, infection and other factors that may influence their egg production. Therefore, using genetic selection, geneticists have managed to come up with egg producing breeds that have high feed conversion efficiency, and have high tolerance to extreme environmental temperatures. These efforts have enabled egg producers maintain adequate feed consumptions and high or optimal feed utilization balance especially under the extremely stressful environmental conditions where adequate nutrition supplementation is required.

In summary, there have been great improvements in the fields of quantitative genetics, genomics, and poultry reproductive functions. A lot of information on technology, immune system genetic functions, and physiology has been integrated into poultry breeding aspects. Some of the novel concepts that have been developed through genetics selection to enhance egg production in poultry industry include disease resistance breeding strains, heat-shock and heat stress proteins, epigenetics, reproductive aging, and other behaviour responses like moulting in egg laying chickens. Geneticists in poultry breeding and production companies have and will continue incorporating novel breeding and production technologies which are beneficial and most appropriate in various genetic selection poultry programs. Improved biological systems for egg layers will continue receiving more and more emphases in order to be able to cope with various challenging climatic conditions, husbandry and feeding practices, and housing systems. Progress in genetic selection for improved egg production and other related characteristics will be realized in order to enhance egg production to meet rapidly increasing producers and consumers demands.

CONCLUSION

Poultry breeders and producers have consistently been using genetic selection in their breeding and production programs in order to improve performance in their stocks. The main focus has always been to enhance feed consumption and how those feeds are converted efficiently into eggs and meat. Geneticists have managed to reduce waste production in broiler chicken rearing using improved feed efficiency strategies.

Through genetic selection, the survival of broiler chickens has been improved amicably. Different novel traits have also been incorporated genetically into various commercial selection programs that are used in helping further enhance legs and skeletal health of chickens, integrity of the skin, fitness of cardiovascular, and resistance in disease in broiler and layers.

More genomic and genetic studies are underway in effort to develop different selection tools for welfare and health of the birds (broiler and layers) including human related food safety bacterial resistance and enhanced inborn immunity in poultry.

Poultry breeding for enhanced health and welfare of the birds and for future sustainability need continued addition and incorporation of more operational and accurate measurements on various traits plus novel application of genomics and molecular genetics in order to help in solving various problems experienced in efficiency and production of eggs and meat.

REFERENCES

1. Arthur, J. A., & Albers, G. A. (2003). Industrial perspective on problems and issues associated with poultry breeding. *Poultry genetics, breeding and biotechnology*, 1-12.
2. Blanchard, J. R. (1952). *Literature of Agricultural Research.* California: University of California Press.
3. Daghir, N. J. (2008). *Poultry Production in Hot Climates.* Wallingford: CABI.
4. Dan Phillip Sponenberg, Donald E. Bixby. (2007). *Managing Breeds for a Secure Future: Strategies for Breeders and Breed Associations.* London: ALBC.
5. Donald D. Bell, William D. Weaver. (2012). *Commercial Chicken Meat and Egg Production.* London: Springer Science & Business Media.
6. Eiri. (2007). *Hand Book Of Poultry Farming And Feed Formulations.* India: Engineers India Research In.
7. Emmanuel Babafunso Sonaiya, S. E. J. Swan. (2004). *Small-scale Poultry Production: Technical Guide.* London: Food & Agriculture Org.

8. Fair-f&, R. W., L. M'Millan, and W. M. Mui. (1998). Poultry breeding: Progress and prospects for genetic improvement of cgg and meat production.
9. Flock, D. K., K. F. Laughlin, and J. Bentley. (2005). Minimizing losses in poultry breeding and production: how breeding companies contribute to poultry welfare. *World's Poultry Science Journal* , 227-237.
10. Frank Flanders, James R. Gillespie. (2015). *Modern Livestock & Poultry Production.* New York: Cengage Learning.
11. Fulton, J. E. (2008). Molecular genetics in a modern poultry breeding organization. *World's Poultry Science Journal*, 171-176.
12. Fulton, J. E. (2012). Genomic selection for poultry breeding. *Animal Frontiers*, 30-36.
13. Fulton, J. E., & Delany, M. E. . (2003). Poultry genetic resources--operation rescue needed. *Science*, 300(5626), 1667-1668.
14. Hocking, P. M. (2009). *Biology of Breeding Poultry.* Wallingford: CABI.
15. Hutt, F. B. (2003). *Genetics of the Fowl: The Classic Guide to Chicken Genetics and Poultry Breeding.* Michiga: Norton Creek Press.
16. Institute of Laboratory Animal Resources (U.S.). Subcommittee on Avian Standards. (1966). *Standards and Guidelines for the Breeding, Care, and Management of Laboratory Animals: Chickens; a Report.* London: National Academies.
17. Jaap, R. George, and W. F. Hollander. (1956). Wild type as standard in poultry genetics. *Poultry Science*, 94-100.
18. Pym, Robert. (2013). Poultry genetics and breeding in developing countries. *the role of poultry in human nutrition*, 80.
19. R. Gous, C. Fisher, Trevor Raymond Morris. (2006). *Mechanistic Modelling in Pig and Poultry Production.* Wallingford: CABI.
20. Rowen D. Frandson, W. Lee Wilke, Anna Dee Fails. (2009). *Anatomy and Physiology of Farm Animals.* Chicago: John Wiley & Sons.
21. Ruth Bellairs, Mark Osmond. (2005). *Atlas of Chick Development.* London: Academic Press.

22. Victoria Sandilands, Paul M. Hocking. (2012). *Alternative Systems for Poultry: Health, Welfare and Productivity.* Wallingford: CABI.
23. W. M. Muir, S. E. Aggrey. (2003). *Poultry Genetics, Breeding, and Biotechnology.* Wallingford: CABI.

INDEX